ENGINEERING BULLETIN

Concrete Floors on Ground

by James A. Farny

PCA PORTLAND CEMENT ASSOCIATION

An organization of cement companies to improve and extend the uses of portland cement and concrete through market development, engineering, research, education, and public affairs work.

5420 Old Orchard Road, Skokie, Illinois 60077-1083 USA

About the author: The author of this engineering bulletin is James A. Farny, Program Coordinator, Masonry and Special Products, Portland Cement Association.

On the cover: Illustrations on the cover are clockwise from top left (1) a ride-on trowel (69871); (2) high rack warehouse floor (69655); and (3) placing concrete for a floor with a pump and a laser screed (69718).

Third Edition

Print History:
First Printing, 2001
Second Printing, 2006

PHOTO AND FIGURE CREDITS

Several individuals and companies shared photographs and other figures for this publication. They include: Baker Concrete Construction, Inc.; The Euclid Chemical Company; Howard Kanare and Scott Tarr, CTL; John Rohrer Contracting Company; Hanley-Wood, LLC (*Concrete Construction* magazine); Lehigh Portland Cement Company; Maxxon Corporation; National Ready Mixed Concrete Association; and Peter Nussbaum, formerly of CTL.

ISBN 0-89312-211-4

PCA R&D Serial No. 2524

Printed in the United States of America

EB075.03

ACKNOWLEDGMENTS

Many people were involved with bringing this technical handbook to print. Not all of them can be mentioned here, but several individuals should be recognized. This document is an updated and expanded version of Concrete Floors on Ground (EB075.02D), by Ralph E. Spears (deceased) and William C. Panarese, respectively, former manager, Building Construction Section, and former manager, Construction Information Services, PCA. Peter Nussbaum, formerly of CTL, did the initial review of the previous edition, suggested many changes, and provided a final technical review of the new edition. Two notable experts shared their knowledge to ensure that the best current practice is provided: Steven Metzger, consultant, Metzger-McGuire, on jointing, joint filling and sealing, and joint repair; and Howard Kanare, Construction Technology Laboratories, Inc., on moisture in concrete and moisture testing. Overall technical reviews were given by Terry Collins, PCA's Concrete Construction Engineer, Martin McGovern, PCA's Concrete Engineer, Warren Baas of Ohio Ready Mix, and Carl Bimel, consultant and chairman of ACI's Committee 302, Slabs on Ground. Steven Kosmatka, PCA's Managing Director of Research and Technical Services, provided assistance throughout the entire revision. Thank you all.

Warning: Contact with wet (unhardened) concrete, mortar, cement, or cement mixtures can cause SKIN IRRITATION, SEVERE CHEMICAL BURNS (THIRD-DEGREE), or SERIOUS EYE DAMAGE. Frequent exposure may be associated with irritant and/or allergic contact dermatitis. Wear waterproof gloves, a long-sleeved shirt, full-length trousers, and proper eye protection when working with these materials. If you have to stand in wet concrete, use waterproof boots that are high enough to keep concrete from flowing into them. Wash wet concrete, mortar, cement, or cement mixtures from your skin immediately. Flush eyes with clean water immediately after contact. Indirect contact through clothing can be as serious as direct contact, so promptly rinse out wet concrete, mortar, cement, or cement mixtures from clothing. Seek immediate medical attention if you have persistent or severe discomfort.

Concrete Floors on Ground

TABLE OF CONTENTS

CHAPTER

CHAPTER

CHAPTER

CHAPTER

CHAPTER

CHAPTER

CHAPTER 1

CONCRETE FLOORS ON GROUND

TYPES OF FLOORS

A concrete floor on ground is a common element of concrete construction (see Fig. 1-1). It can be a simple slab or very complex. Table 1-1 describes nine classes of floors put forth by the American Concrete Institute. No matter how basic or highly engineered any floor is, the construction method is similar: ground is prepared and concrete is placed. Of course, there are many other considerations such as drainage and thickness design. Table 1-1 notes some of the special considerations and finishing techniques that are appropriate for each class of floor.

Fig. 1-1. There are many applications for concrete floors on ground, including industrial facilities, warehouses, basements, and garages. (69648, 69610, 37459, 69650)

Table 1-1. Floor Classifications

Class	Anticipated type of traffic	Use	Special considerations	Final finish
1 Single course	Exposed surface–foot traffic	Offices, churches, commercial, institutional, multi-unit residential Decorative	Uniform finish, non-slip aggregate in specific areas, curing Colored mineral aggregate, color pigment or exposed aggregate, stamped or inlaid patterns, artistic joint layout, curing	Normal steel-troweled finish, non-slip finish where required As required
2 Single course	Covered surface–foot traffic	Offices, churches, commercial, gymnasiums, multi-unit residential, institutional with floor coverings	Flat and level slabs suitable for applied coverings, curing Coordinate joints with applied coverings	Light steel-troweled finish
3 Two course	Exposed or covered surface–foot traffic	Unbonded or bonded topping over base slab for commercial or non-industrial buildings where construction type or schedule dictates	*Base slab*—good, uniform, level surface, curing *Unbonded topping*—bond-breaker on base slab, minimum thickness 75 mm (3 in.) reinforced, curing *Bonded topping*—properly sized aggregate, 19 mm (3/4 in.) minimum thickness, curing	*Base slab*—troweled finish under unbonded topping; clean, textured surface under bonded topping *Topping*—for exposed surface, normal steel-troweled finish, for covered surface, light steel-troweled finish
4 Single course	Exposed or covered surface–foot and light vehicular traffic	Institutional and commercial	Level and flat slab suitable for applied coverings, non-slip aggregate for specific areas, curing. Coordinate joints with applied coverings	Normal steel-troweled finish
5 Single course	Exposed surface–industrial vehicular traffic (pneumatic wheels, and moderately soft solid wheels)	Industrial floors for manufacturing, processing, and warehousing	Good uniform subgrade, joint layout, abrasion resistance, curing	Hard steel-troweled finish
6 Single course	Exposed surface-heavy duty industrial vehicular traffic (hard wheels, and heavy wheel loads)	Industrial floors subject to heavy traffic; may be subject to impact loads	Good uniform subgrade, joint layout, load transfer, abrasion resistance, curing	Special metallic or mineral aggregate surface hardener; repeated hard steel troweling
7 Two course	Exposed surface–heavy duty industrial vehicular traffic (hard wheels, and heavy wheel loads)	Bonded two-course floors subject to heavy traffic and impact	*Base slab*—good, uniform subgrade, reinforcement, joint layout, level surface, curing *Topping*—composed of well-graded all-mineral or all-metallic aggregate. Minimum thickness 19 mm (3/4 in.). Metallic or mineral aggregate surface hardener applied to high-strength plain topping to toughen, curing	Clean, textured base slab surface suitable for subsequent bonded topping. Special power floats for topping are optional, hard steel-troweled finish
8 Two course	As in Class 4, 5, or 6	Unbonded toppings on new or old floors or where construction sequence or schedule dictates	Bondbreaker on base slab, minimum thickness 100 mm (4 in.), abrasion resistance, curing	As in Class 4, 5, or 6
9 Single course or topping	Exposed surface–superflat or critical surface tolerance required. Special materials-handling vehicles or robotics requiring specific tolerances	Narrow-aisle, high-bay warehouses; television studios, ice rinks	Varying concrete quality requirements. Shake-on hardeners cannot be used unless special application and great care are employed. F_F 50 to F_F 125 ("superflat" floor), curing	Strictly follow finishing techniques as indicated for superflat floors

Adapted from American Concrete Institute's ACI 302.1R, *Guide for Concrete Floor and Slab Construction.*

Floor construction should be straightforward. This text has been prepared to help designers and builders gain the knowledge to create long-lasting floors that perform the functions for which they were designed.

Large-area concrete floors for commercial and industrial buildings must be designed and constructed with the greatest possible economy to give trouble-free service year after year. The building of a good concrete floor requires close communication between owner, architect, engineer, and contractor—with a mutual understanding of the level of quality needed for its intended use.

FLOOR PERFORMANCE CRITERIA

Anticipated floor service conditions and the criteria used to measure floor performance will determine the design sophistication, type of materials, and workmanship needed. Prior to bidding, all parties—owners, users, designers, contractors and subcontractors—must know about and agree upon these conditions and criteria.

A good concrete floor on ground is the result of many factors:

- sensible planning
- careful design and detailing
- proper materials selection
- complete specifications
- proper inspection
- good workmanship

To define the responsibilities of each participant, pre-design, pre-bidding, and pre-construction meetings are essential. At the pre-design conference, owners and users should answer several questions.

Fig. 1-2. High rack storage requires very flat floor conditions for proper operation of lift truck equipment. (69649)

- How will the floor be used?
- What types and magnitudes of floor loadings are anticipated?
- What are the aesthetic requirements, including acceptability of random slab cracking?
- Will housekeeping procedures for the floor influence how the floor is built?
- Will floor coverings, coatings, or special toppings be used?

During the pre-bid meeting, special concerns or unusual requirements for the proposed construction should be discussed. Potential bidders should have the opportunity to clarify any questions they have about the bid documents.

The pre-construction meeting and in-progress meetings during construction provide a forum for communication between contractors, subcontractors, quality assurance staff, and materials suppliers. In-progress addenda to plans and specifications should be documented. Deficiencies in workmanship or in the contract documents should be discussed. Actions to resolve problems should be documented.

The construction team includes the owner, designer, general contractor, subcontractors concerned with subgrades and subbases, concrete supplier, concrete contractor, and quality assurance engineers. The entire team is responsible for producing a long-term, trouble-free floor.

Once the operation criteria are defined and incorporated into the floor design and specifications, it is up to the construction team to deliver the required floor.

Many concrete floors built today are highly engineered structures with significant demands placed on them. For example, new warehouse facilities are often designed for high-density storage. This means the facility will have tall shelving, will have to carry extremely heavy loads, and will need very flat and level surfaces for smooth operation of the tall lift trucks that store and move goods (see Fig. 1-2). Such a highly engineered floor would not be necessary, or cost effective, for homes or small businesses.

Historically perceived as simple elements, floors often received little attention during the building process. In addition, a number of incorrect ideas developed about the effect of good concrete practices on the serviceability of floors on ground. As a result, floors probably account for more complaints than any other part of a building. To correct this deficiency, careful attention must be given to a number of factors that influence floor performance:

- uniformity of the subgrade and adequacy of its bearing capacity
- quality of the concrete
- adequacy of structural capacity (thickness)
- surface levelness and flatness
- deformations under load
- load transfer at joints
- type and spacing of joints
- workmanship

- under-slab treatments
  (vapor retarders, capillary breaks)
- concrete moisture content and drying rate
- special surface finishes, including coatings
- future maintenance and repair

All these factors are covered in detail in this text. The technology and details apply to floors of all sizes, encompassing a wide variety of uses. From small-area floors in a residence or light industry, to medium-sized warehouse floors, to heavy industrial plants with storage facilities that cover large areas, the technology is similar. This book is geared primarily, but not exclusively, toward interior floors. The basic principles also apply to special-use concrete slabs (indoors or outdoors), such as tennis courts, track-and-field facilities, and pool decks, to name a few. Major emphasis is given to attaining the best possible balance between service requirements and costs of construction and future maintenance.

COST EFFECTIVENESS OF THICKER FLOORS

Aside from the common deficiencies of inadequate curing, joints spaced too far apart, and overly wet mixes, the most overlooked but fundamental problem has been slab thickness. Increasing the slab thickness is one of the easiest and most effective ways to improve a floor's performance. Bending strength increases with the square of a member's depth. For instance, compare a beam that is 1 unit thick to one that is 1.5 times as thick:

1 squared is 1
1.5 squared is 2.25

Therefore the bending strength of the thicker beam would be 2.25/1, or 225% of the thinner beam. Slabs follow the same type of relationship. To illustrate, compare a 100-mm (4-in.) unreinforced floor slab on grade with one that is 125 mm (5 in.) thick. To make the illustration generic, the comparison can be based on percentages. Let the cost of the concrete in the 100-mm (4-in.) slab be 100%. The cost of additional material in the 125-mm (5-in.) slab is 25% more, or a total of 125%. (The actual cost in place of the thicker slab, including labor, would be a lower percentage, because subgrade preparation and surface finishing costs would be the same.)

To compare the bending strength of the 125-mm (5-in.) slab with the 100-mm (4-in.) slab, consider the ratio of 1252 to 1002 (or 252 to 42), which equals 15,625/10,000 (625/16), or 1.56. In this example, for a 25% increase in material cost, there is a 56% increase in load-carrying capacity. Similarly, for a 100-mm (4-in.) slab that is increased 50% in thickness, there is a 125% increase in load-carrying capacity. *For every dollar that is spent to produce a thicker floor, there is more than twice as much value in actual load-carrying capacity.* Table 1-2 shows the relative improvement in load-carrying capacity resulting from thickening the slab. A similar relationship applies when a slab of any thickness is increased in depth.

Table 1-2. Slab Thickness and Load-Carrying Capacity

Slab thickness, mm (in.)	Relative Thickness compared to a 100-mm (4-in.) slab	Relative strength compared to a 100-mm (4-in.) slab
100 (4)	100	100
125 (5)	125	156
150 (6)	150	225
175 (7)	175	306
200 (8)	200	400

CHAPTER 2

SUBGRADE AND SUBBASES

To ensure that the concrete floor will carry its design loading successfully and without settlement, it is vital to design and construct the subgrade in preparation for the floor. A subbase, while not mandatory, can provide added benefits in construction and performance (see Fig. 2-1).

SUBGRADE CHARACTERISTICS AND FUNCTION

The subgrade is the natural ground, graded and compacted, on which the floor is built. The subgrade can be improved by drainage, compaction (see "Density"), or soil stabilization. In cases of extremely poor soil, removal and replacement of the subgrade with a compactible material may be the best option. Subgrade support should be reasonably uniform without abrupt changes from hard to soft, and the upper portion of the subgrade must be of uniform material and density.

Because concrete floor slabs are rigid, concentrated loads from forklift wheels or high-rack legs are spread over large areas and pressures on the subgrade are usually low. Therefore, concrete floors do not necessarily require strong support from the subgrade. However, support from subgrades and/or subbases contributes to solid edge support, which is beneficial at slab joints exposed to heavy lift-truck loadings. If support from the subgrade or subbase is weak, densification may occur as heavy loads pass over the floor, leading to loss of slab edge support.

In some instances, pressures transmitted to subgrade soils can be significant. This is true where heavy goods—like steel products or large rolls of paper—are stored directly on the floor (see Fig. 2-2). In these cases, long-term soil consolidation effects should be considered in subgrade performance.

SOILS

Classification

Proper classification of the subgrade soil must be made to identify potential problem soils (see *PCA Soil Primer*, EB007). One soil classification system in common use is

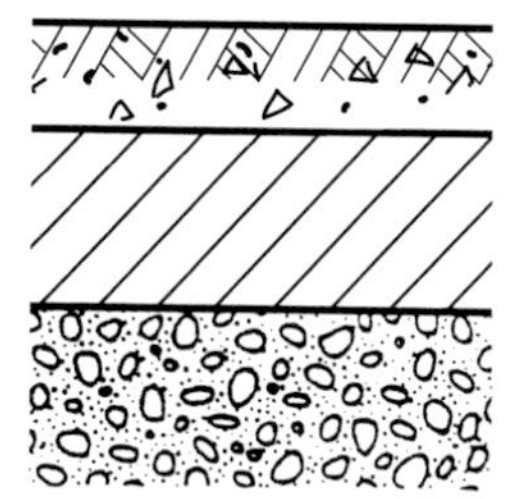

1. Wearing surface: Portland cement concrete, typically a thickness of 100 mm (4 in.) or greater.
2. Subbase material (optional): Compactible, drainable fill. Typical thickness is 75 mm – 100 mm (3 in. – 4 in.) and a fine material may be spread over the surface to fill voids (choker material), which reduces friction with concrete topping.
3. Subgrade: The natural soil at the site. The top 150 mm – 500 mm (6 in. – 20 in.) is usually compacted prior to the placement of the other layers of the slab.

Adapted from Holtz and Kovacs, *Introduction to Geotechnical Engineering*.

Fig. 2-1. (top) This cross-section shows the general relationship between the subgrade, the subbase, and the slab. (middle-69422) Here is a prepared subgrade ready for placement of subbase material. (bottom-69421)

Fig. 2-2. Heavy loads on a concrete floor. (69651)

Classification of Soils for Engineering Purposes, American Society for Testing and Materials (ASTM) Designation D 2487. Table 2-1, based on the ASTM system, shows the major divisions of soils with descriptive names and letter symbols indicating their principal characteristics. When soil is a combination of two types, it is described by combining both names. Thus, clayey sand is predominantly sand but contains an appreciable amount of clay. Reverse the name to sandy clay and the soil is predominantly clay with an appreciable amount of sand.

Density

The strength of the soil—its supporting capacity and resistance to densification or consolidation—is important to the performance of floors on ground, particularly when the floor must support extremely heavy loads. Soil strength is affected by soil type, degree of compaction, and moisture content.

Compaction is a method for purposely densifying or increasing the unit weight of a soil mass by rolling, tamping, or vibrating. It is the lowest-cost way to improve the structural properties of a soil. Density of a soil is measured in terms of its mass per unit volume. Higher densities usually provide improved support.

Tests performed according to *Moisture-Density Relations of Soils,* ASTM D 698 and D 1557, will determine the maximum density and corresponding optimum moisture content of the soil (see Fig. 2-3). Soil moisture contents are expressed in percent: the ratio of the mass of water divided by the mass of dry soil, then multiplied by 100.

Plasticity Index

When a soil can be rolled into thin threads, it is called plastic. Most fine-grained soils containing clay minerals are plastic. The plastic limit (PL) is the amount of moisture present when a soil changes from semisolid to plastic state. The liquid limit (LL) is the amount of moisture present when the soil changes from a plastic to liquid state. The degree of plasticity is expressed as the plasticity index (PI). The PI is the numerical difference between the liquid limit and plastic limit.

$$PI = LL - PL$$

Problem Soils

Soils are considered problem soils when they:

- are highly expansive
- are highly compressible
- do not provide reasonably uniform support

Cohesive soils that gain moisture after being compacted (for example, by a rising water table) can cause problems for concrete floors. Soils with a PI of 5 or greater can deform plastically when the soil moisture content increases. If this happens, loads passing over the floor may lead to deflections due to the loss of bearing support strength. Sensitive floor locations are near slab edges and corners exposed to repetitive wheeled traffic. Cohesive soils with a high PI (greater than 20) that gain soil moisture may become expansive, putting excessive pressure on the slab from below.

Concrete-floor-on-ground design is based on the assumption of uniform subgrade support. The key word is "uniform." Where problem soils create nonuniform conditions, correction is most economically and effectively achieved through subgrade preparation methods.

SITE PREPARATION

To construct a reasonably uniform subgrade, special care must be taken to ensure that there are no variations of support within the floor area and that the following major causes of nonuniform support are controlled:

- expansive soils
- hard spots and soft spots
- backfilling

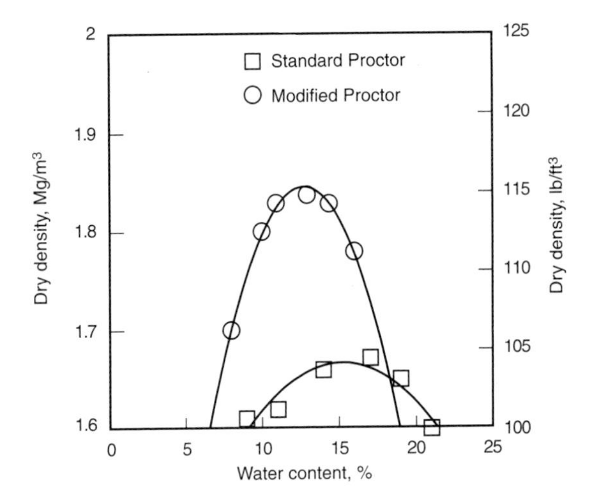

Fig. 2-3. Soil testing determines the optimum moisture concrete that allows for maximum compaction.

Table 2-1. ASTM Soil Classification System

Major divisions			Group symbols	Typical names	Presumptive bearing capacity,** tons per square foot (MPa)	Modulus of subgrade reaction, pounds per square inch per inch (MPa/m)
Coarse-grained soils moire than 50% retained on 75-μm (No. 200) sieve*	Gravels 50% or more of coarse fraction retained on 4.75-mm (No. 4) sieve	Clean gravels	GW	Well-graded gravels and gravel-sand mixtures, little or no fines	5 (0.48)	300 or more (81 or more)
			GP	Poorly graded gravels and gravel-sand mixtures, little or no fines	5 (0.48)	300 or more (81 or more)
		Gravels with fines	GM	Silty gravels, gravel-sand-silt mixtures	2.5 (0.24)	—
			GC	Clayey gravels, gravel-sand-clay mixtures	2 (0.20)	200 to 300 (54 to 81)
	Sands more than 50% of coarse fraction passes 4.75-mm (No. 4) sieve	Clean sands	SW	Well-graded sands and gravelly, sands, little or no fines	3.75 (0.36)	200 to 300 (54 to 81)
			SP	Poorly graded sands and gravelly sands, little or no fines	3 (0.29)	200 to 300 (54 to 81)
		Sands with fines	SM	Silty sands, sand-silt mixtures	2 (0.20)	200 to 300 (54 to 81)
			SC	Clayey sands, sand-clay mixtures	2 (0.20)	200 to 300 (54 to 81)
Fine-grained soils 50% or more passes 75-μm (No. 200) sieve*	Silts and clays liquid limit 50% or less		ML	Inorganic silts, very fine sands, rock flour, silty or clayey fine sands	1 (0.10)	100 to 200 (27 to 54)
			CL	Inorganic clays of low to medium plasticity, gravelly clays, sandy clays, silty clays, lean clays	1 (0.10)	100 to 200 (27 to 54)
			OL	Organic silts and organic silty clays of low plasticity	—	100 to 200 (27 to 54)
	Silts and clays liquid limit greater than 50%		MH	Inorganic silts, micaceous or diatomaceous fine sands or silts, elastic silts	1 (0.10)	100 to 200 (27 to 54)
			CH	Inorganic clays of high plasticity, fat clays	1 (0.10)	50 to 100 (13.5 to 27)
			OH	Organic clays of medium to high plasticity	—	50 to 100 (13.5 to 27)
Highly organic soils			PT	Peat, muck, and other highly organic soils	—	—

* Based on the material passing the 75-mm (3-in.) sieve.
** *National Building Code.* 1976 Edition. American Insurance Association.

Table 2-2. Expansion Versus Plasticity

Degree of expansion	Percentage of swell	Approximate plasticity index (PI)
Nonexpansive	2 or less	0 to 10
Moderately expansive	2 to 4	10 to 20
Highly expansive	more than 4	more than 20

Expansive Soils

Most soils sufficiently expansive to cause floor distortion are classified by the ASTM Soil Classification System (Table 2-1) as clays of high plasticity (CH), silts of high plasticity (MH), and organic clays (OH). Simple soil tests provide indexes that serve as useful guides to identify the approximate volume-change potential of soils. (Table 2-2 shows approximate expansion-plasticity relationships.)

Following are some of the soil tests:

Test Method for Liquid Limit, Plastic Limit, and Plasticity Index of Soils, ASTM D 4318

Test Method for CBR (California Bearing Ratio) of Laboratory-Compacted Soils, ASTM D 1883

Test Method for Shrinkage Factors of Soils by Mercury Method, ASTM D 427

Test Method for Laboratory Compaction Characteristics of Soil Using Standard Effort (12,400 ft-lbf/ft³ (600 kN-m/m³)), ASTM D 698- also known as the "Standard Proctor Test"

Test Method for Laboratory Compaction Characteristics of Soil Using Modified Effort (56,000ft-lbf/ft³ (2,700 kN-m/m³)), ASTM D 1557 also known as the "Modified Proctor Test"

Abnormal shrinkage and swelling of high-volume-change soils in a subgrade will create nonuniform support. As a result, the concrete floor may become distorted. Compaction of highly expansive soils when the soils are too dry can contribute to detrimental expansion and softening of the subgrade upon future wetting. When expansive soil subgrades are too wet prior to casting a floor slab, subsequent drying and shrinkage of the soil may leave portions of the slab unsupported.

Selective grading, crosshauling, and blending of subgrade soils make it possible to obtain uniform conditions in the upper part of the subgrade. Compaction of expansive soils will minimize possible loss of support from any future increases in moisture content and give the subgrade the uniform stability that is needed for good performance (see Fig. 2-4). As a general rule, compaction to 95% optimum density at 1% to 3% above standard optimum moisture (Soil Compaction Tests, ASTM D 698 and D 1557) will stabilize expansive soils.

For very heavy loadings or poor soil conditions, an investigation by a soils engineer is warranted. The soils engineer should provide data on subgrade soil shear strength and soil consolidation to be used for bearing capacity and settlement calculations.

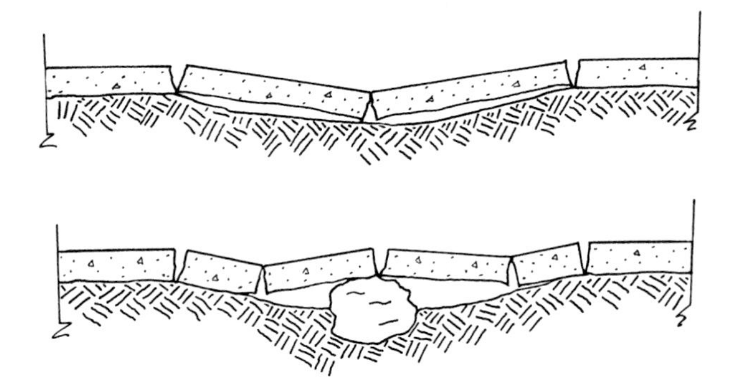

Fig. 2-4. Soft spots and hard spots.

Hard Spots and Soft Spots

If the subgrade provides nonuniform support, the slab when loaded will tend to bridge over soft spots and ride on hard spots, all too often with the results illustrated in Fig. 2-4. Special care must be taken when excavating to prevent localized soft or hard spots. Uniform support, however, cannot be obtained merely by dumping granular material on the soft spot. Moisture and density conditions of the replacement soil should be as similar as possible to the adjacent soil. At transition areas where soil types or conditions change abruptly, the replacement soil should be mixed with the surrounding soil by crosshauling and blending to form a transition zone with uniform support conditions.

Backfilling

Any fill material added to improve the subgrade or raise the existing grade should be a stable material that can be thoroughly compacted. Rubble from building or pavement demolitions must first be passed through a crusher because large pieces can cause compaction difficulties.

Backfilling at footings, foundations, and pipeline and utility trenches should be done with soils like those surrounding the trench and compacted in layers to duplicate moisture and density conditions in the adjacent soils. Every attempt should be made to restore as much as possible the in-place uniformity of the compacted subgrade. For clays, compacted layer thickness should not exceed about 150 mm (6 in.). When backfilling with granular material, place the material in layers and thoroughly compact it with vibratory equipment.

Poorly compacted subgrade fill can cause subsequent settlement problems and loss of slab support, resulting in premature failure of the slab.

Recently, controlled low-strength material (CLSM) has been used as an easily placed fill material. CLSM is a self-compacted cementitious material used primarily for backfill as an alternative to compacted fill. Controlled low-strength materials have a compressive strength of 8.3 MPa (1200 psi) or less. They are also known as flowable fills. CLSM can be used to bring a foundation up to proper grade prior to placing the concrete floor slab. CLSM is especially useful where there are a lot of obstructions like spread footings and hand compaction of select granular backfill would be difficult, time consuming, and expensive. ACI 229R, *Controlled Low-Strength Materials,* covers important mix design information and construction techniques, helpful to the success of projects incorporating CLSM.

SUBGRADE-SUBBASE STRENGTH

Soil bearing capacity, soil compressibility, and soil modulus of subgrade reaction are various measures of strength-deformation properties of soil. It is important to consider how these parameters apply to the design of floor slabs.

The allowable soil bearing capacity is the maximum pressure that can be permitted on foundation soil with adequate safety against soil rupture or excessive settlement. Allowable soil pressure may be based on:

- laboratory shear strength tests (of soil samples) such as the direct shear test, triaxial compression test, or unconfined compression test
- field tests such as the standard penetration test or cone penetrometer test
- soil classification
- moisture-density-strength relationships (established by conducting strength tests on soil specimens prepared for moisture-density testing)

Beyond the allowable soil pressure is the ultimate bearing capacity, the load per unit area (soil pressure) that will produce failure by rupture of a supporting soil.

Another soil characteristic, compressibility of cohesive soils, determines the amount of long-term settlement under load. The usual method for predicting settlement is based on conducting soil consolidation tests and determining the compression index for use in the settlement computations. The compression index may be estimated by correlation to the liquid limit of the soil.

A third measure of soil strength, Westergaard's modulus of subgrade reaction, *k*, is commonly used in design procedures for concrete pavements and floors-on-grade that are not structural elements in the building (floors not supporting columns and load-bearing walls).

There is no reliable correlation between the three measures of soil properties—modulus of subgrade reaction, soil bearing capacity, and soil compressibility—because they are measurements of entirely different characteristics of a soil. The *k*-value used for floor-slab design reflects the response of the subgrade under temporary (elastic) conditions and small deflections, usually 1.25 mm (0.05 in.) or less. Soil compressibility and bearing capacity values (normally used to predict and limit differential settlements between footings or parts of a foundation) reflect total permanent (inelastic) subgrade deformations that may be 20 to 40 (or more) times greater than the small deflections on which *k*-values are based.

Substantial pavement research has shown that elastic deflections and stresses of the slab can be predicted reasonably well when using *k*-value to represent the subgrade response. Consequently, the control of slab stresses based on the subgrade *k*-value is a valid design procedure.

Although the *k*-value does not reflect the effect of compressible soil layers at some depth in the subgrade, it is the correct factor to use in design for wheel loads and other concentrated loads because soil pressures under a slab of adequate thickness are not excessive. However, if heavy distributed loads will be applied to the floor, the allowable soil pressure and the amount of settlement should be estimated to determine if shear failure or excessive settlement might occur.

If there are no unusually adverse soil conditions, the design analysis requires only the determination of the strength of the subgrade in terms of *k*. The *k*-value is measured by plate-loading tests taken on top of the compacted subgrade (or subbase, if used) as seen in Fig. 2-5. A general procedure for load testing is given in ASTM D 1196, *Standard Test Method for Nonrepetitive Static Plate Load Tests of Soils and Flexible Pavement Components, for Use in Evaluation and Design of Airport and Highway Pavements.* This method provides guidance in the field determination of subgrade modulus with various plate diameters. *Design of Slabs on Grade* (ACI 360R) is specifically oriented to the determination of modulus of subgrade reaction using a 760-mm (30-in.) diameter plate and gives more detailed information on test methods using this size plate. This plate is loaded to a deflection not greater than 1.25 mm (0.05 in.), and the *k*-value is computed by dividing the unit load by the deflection obtained. A more economical test using smaller plates (300 mm [12 in.]) that determines a modified subgrade reaction modulus is mentioned in ACI 360R. In each case, the units of *k* are given in pressure per length: MPa/m in the metric system, or in in.-lb units, pounds per square inch per inch, or psi per in. or, as commonly expressed, pounds per cubic inch (pci). The plate load test is no longer commonly run in practice. Instead, subgrade reaction values are estimated from the California Bearing Ratio or from the soil classification. When plate-bearing tests are not performed at the jobsite, the *k*-value can be estimated from correlations such as those shown in Table 2-3.

Fig. 2-5. Testing to determine the modulus of subgrade reaction. (69652)

Table 2-3. Estimating the Subgrade Modulus

Type of soil	Subgrade strength	CBR,[2] percent	Design *k*-value	
			pci	MPa/m
Silts and clays of high compressibility[1] at natural density	Low	2 or less	50	13.6
Silts and clays of high compressibility[1] at compacted density Silts and clays of low compressibility[1] Sandy silts and clays, gravelly silts and clays Poorly-graded sands	Average	3	100	27.1
Gravelly soils, well-graded sands, and sand-gravel mixtures relatively free of plastic fines	High	10	200	54

[1]High compressibility, liquid limit equal to or greater than 50. Low compressibility, liquid limit less than 50. (Liquid limit by ASTM D 4318, *Test Method for Liquid Limit, Plastic Limit, and Plasticity Index of Soils.*)
[2]California Bearing Ratio, ASTM D 1883, *Test Method for CBR (California Bearing Ratio) of Laboratory-Compacted Soils.*

If a high-quality, well-compacted granular subbase is used under the floor slab, the *k*-value will increase. On large projects it may be feasible to construct a test section and perform plate load tests on top of the subbase. If this is not practical, the *k*-value on top of the subbase can be estimated from Fig. 2-6.

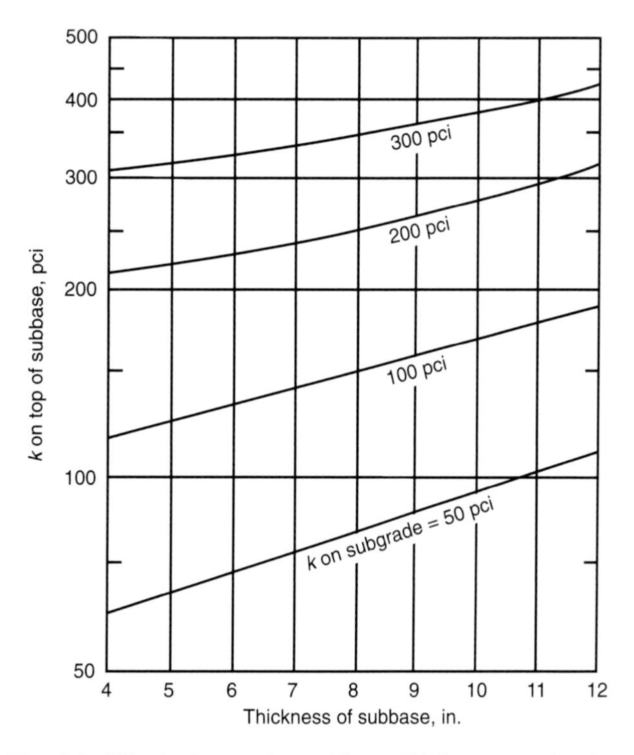

Fig. 2-6. Effect of granular subbase thickness on *k*-value. (1 pci = 0.27 MPa/m)

SUBBASES

A subbase—the layer of granular material placed on top of prepared subgrade—is not mandatory for floors on ground. A granular subbase, however, can provide benefits during the construction process and afterwards (to the completed floor). During construction, the subbase functions as a stable work platform for heavy equipment. When grading and compaction operations do not produce a uniform subgrade, a granular subbase will provide a cushion for more uniform slab support by equalizing minor subgrade defects. The cushioning effect and increased uniformity can be very important for cohesive soils that are susceptible to reduced bearing support with increases in moisture content. A subbase also serves as a capillary break, reducing moisture migration towards the bottom of the completed slab. A granular subbase can also serve as a collection layer for radon gas (see box on radon, page 9).

Contributions to modulus of subgrade reaction, *k*, beyond 100-mm (4-in.) subbase thickness are illustrated in Fig. 2-6. In terms of floor slab thickness design, effects of increased *k* are most significant for the lower subgrade support values (as indicated by the steeper slope for smaller *k*s in Fig. 2-6). In other words, floor thickness for a given loading condition is not significantly reduced by increased subbase thickness when stronger subgrades are located below the subbase. Thus, it is seldom necessary or economical to build up the supporting capacity of the subgrade with a thick subbase.

Since uniform rather than strong support is the most important function of the subgrade and subbase for a concrete floor, it follows that floor strength is achieved most economically by building strength into the concrete slab itself—with optimum use of low-cost materials under the slab.

Radon

Radon is an odorless, colorless, radioactive gas. Radon is produced naturally when uranium and radium in the soil decay. Though radon exists virtually everywhere, the concentration varies with geographic location. Parts of Pennsylvania, New Jersey, New York, Florida, Colorado, and Montana have been identified as having high radon concentrations. It is most prevalent in highly permeable soil with high concentrations of uranium or radium.

Soil gas enters a building through openings in the foundation walls or slab. Soil gas can have a high radon content. Radon gas itself breaks down into solid particles called progeny, or "daughters," which can lodge in lung tissue and increase the chances of lung cancer.

Radon is measured in picocuries per liter of air, or pCi/L, a measure of the radioactivity. In 1988, The Indoor Radon Abatement Act established a long-term goal of achieving radon levels inside a building no higher than the ambient air outside the building. Test kits are used to measure the average concentration of radon in the air. In the U. S., the average concentration of radon is 1.2-1.5 pCi/L. When the level exceeds 4 pCi/L, action should be taken to reduce radon gas concentration.

Typical features of a high-risk house include:

- high concentration of radium in the soil beneath the house
- a sump opening
- exposed dirt or wood floor in the basement
- cracks in the foundation or basement slab
- well water

Where radon is a potential concern for residential buildings, proper design and construction are necessary to minimize the entrance of soil gas into the living space. A simple method for controlling indoor radon is to collect the gas, then vent it to the outside. In many instances, a 100-mm to 150-mm (4-in. to 6-in.) thick layer of a compacted aggregate placed beneath the slab is an acceptable method of collection. Coarse aggregate contains sufficient voids to allow for adequate air movement. If sand is used, a geotextile drainage mat will increase the air movement above the collection layer.

A polyethylene sheet placed on top of the aggregate will seal the surface and help contain the gas: it is known as a soil gas retarder. Sheets for soil gas retarders can be 0.15 mm (6 mil) thick, but if they also serve as vapor retarders, they may have to be thicker. Sheets should be lapped at joints, fit closely around all penetrations, and sealed where punctures or tears occur.

The concrete for the slab should have minimal shrinkage and cracking. Large holes should be filled with mortar, grout, or foam to create airtight seals. Openings like sumps or drains should be plugged. All cracks and joints should be sealed with elastomeric joint sealants such as polyurethane or concrete-rated silicone.

Before placing the concrete slab, a vertical pipe should be embedded into the subslab aggregate layer and extended up to the rooftop or to an exterior wall to vent gases away from any fresh air intakes (windows, doors). The pipe should be a minimum of 75 mm (3 in.) in diameter and have a tee on the end beneath the slab.

To prevent subbase densification after the floor is in service, subbase material should be compacted to high density. Determine the optimum moisture content for maximum subbase density by running a standard Proctor laboratory compaction test (ASTM D 698). Because the subbase is only 100 mm (4 in.) thick, field density tests do not accurately measure in-place density. Moisture content samples should be taken to ensure that the material is field compacted at optimum moisture content. Field compaction of the subbase should be verified in some manner. Proofrolling is one of the simplest and most practical methods of determining the ability of the full soil support system to take loads (ACI 302 1997). If any rutting or pumping is evident during the procedure, corrective actions should be taken to improve the soil support.

Granular material for the subbase can be sand, sand-gravel, crushed stone, or combinations of these materials. A satisfactory dense-graded material will meet the following requirements:

Maximum particle size:	Not more than 1/3 the subbase thickness
Passing 75µm (No. 200) sieve:	15% maximum by mass of dry material
Plasticity index:	6 maximum
Liquid limit:	25 maximum

More information on preparing the grade is described in *Subgrades and Subbases for Concrete Pavements.*

Special Subbases

Special subbases are used for a variety of reasons. They consist of one of various materials, including crushed stone, recycled concrete, asphalt, lean concrete, CLSM, and insulation boards. They may be placed to improve subgrade support values, expedite construction, or prevent subgrade frost heave in food freezer warehouses or ice rinks (also see Chapter 11, Special Floors).

Subgrade support can be considerably improved by installing either a cement-treated subbase, CLSM, or lean concrete subbase. A subbase can also be installed as a working platform for erection of facility superstructures. Subbases of CLSM provide better thickness control of floor slabs and a firm surface to support chairs that hold steel reinforcement. The solid support helps with adequate steel placement, which is especially important for installations of shrinkage compensating concrete.

Generally, cement-treated and lean concrete sub-bases are 100 mm to 150 mm (4 in. to 6 in.) thick. A modulus of subgrade reaction of 110 MPa/m to 140 MPa/m (400 pci to 500 pci) can be used to calculate the required thickness of floors placed directly on lean concrete, and cement-treated or roller-compacted subbases.

When insulation is used, a modulus of subgrade reaction test (ASTM D 1196) should be run on a mockup of the insulation layer.

CHAPTER 3
MOISTURE CONTROL AND VAPOR RETARDERS

Good quality concrete resists penetration by liquid water, but will allow water vapor to pass through it slowly. Since most floors are not subjected to water under pressure, it is the passage of water by vapor transmission and capillarity that lead to concerns for moisture control. Moisture movement through concrete floors can cause problems when low-permeability coverings, coatings, or water-sensitive adhesives are applied.

For uncovered floors, moisture movement is typically not a problem because surface evaporation is more rapid than water vapor transmission. Water vapor passing through uncovered floors may be a problem, however, when items are stored directly on the floor. One example of this is cardboard boxes, which may become damp and lose their stacking strength.

Moisture movement through covered floors sometimes causes problems in buildings, including:

- warping, buckling, peeling, and staining of floor coverings (see Fig. 3-1)
- deterioration of floor covering adhesives
- debonding of coatings
- odors and other air quality problems
- damage electrical cable systems

Fig. 3-1. A failed floor covering. (68130)

Preventing problems related to moisture movement requires careful attention to detail during building design and construction. Potential moisture sources must be identified and steps taken to limit exposure of the slab to these sources during construction and under service conditions.

MOISTURE SOURCES

Moisture moving through concrete floors may originate from a high water table, a man-made water source such as an irrigation system, or rain that enters the subgrade or subbase during or after construction. Even concrete mixing or curing water may be a moisture source during the first years of a building's use. Different sources of moisture are sometimes distinguished from each other as originating from either natural or artificial sources.

Soil. Moisture problems can occur even if concrete floors are not in contact with liquid water. Depending on the type of subgrade soil, capillarity can draw moisture more than 3 m (10 ft) for clayey soils and 2 m (7 ft) for fine sands. Thus, high or perched water tables are a possible source of floor moisture even though they are well below the floor or its subbase. Differences in temperature in the soil below the slab and the ambient temperature above the slab have a significant impact on moisture vapor transmission through the concrete floor. As discussed in the previous chapter, coarse-grained granular subbase materials can be used as capillary breaks to limit upward moisture movement through soils.

Artificial Sources. Sprinkler or irrigation systems are a possible source of floor moisture problems, even in arid climates. Landscape planting and watering introduce moisture to the ground, and the designer must take into account potential changes in ground moisture conditions when the building is put into service. Especially when combined with poor site drainage, landscape watering can significantly raise the moisture content of the natural soil (subgrade) and subbase beneath a concrete slab. (Butt 1992)

Precipitation. Rain can also raise the moisture content of the subgrade and subbase before a building is enclosed or, if the site is poorly drained, after construction is complete. If a granular layer is placed directly beneath a floor, excessive sprinkling of this layer before concrete placement produces a similar effect.

Concrete Mix Water. Newly placed concrete contains water, some of which will evaporate after the concrete has been cured and allowed to dry. After sealed curing for 7 days, a cubic meter of normal weight concrete made with 320 kg of cement and 160 kg of water would contain about 95 kg of surplus water in the hydrated cement paste and about 18 kg of water in the aggregate pores. (Conversion to in.-lb: After sealed curing for 7 days, a cu yd of concrete made with 540 lb of cement and 270 lb of water would contain about 160 lb of surplus water in the paste and 30 lb in the aggregate pores.) Some of the surplus water in the paste continues to combine with unhydrated cement and some is adsorbed on surfaces of new hydration products. The rest evaporates at a rate dependent on the temperature and relative humidity at the floor surface.

If the concrete is wet cured—wet coverings, continuous sprinkling, or ponding—the amount of surplus moisture will be slightly higher. The latter two curing methods are not recommended for floors to receive low-permeability coverings, because these methods may raise subgrade or subbase moisture contents to undesirable levels.

Chapter 9 discusses the drying of concrete and other concerns related to floor coverings. This chapter focuses on limiting moisture that passes through the concrete slab, such as when and how to use a vapor retarder, and proper subgrade preparation.

CONTROL OF FLOOR MOISTURE MOVEMENT

Preliminary Design

The floor designer must draw on information from several sources before and during the design process. Site inspection and testing by a geotechnical engineering firm will provide some of the information needed to design for control of floor moisture movement. The geotechnical report should include information about:

- soils: types, conditions, and properties
- groundwater: level of water table, movement (hydrology), and seasonal/historical data (flooding)

Information on soil types identifies soils with high silt and clay contents that can cause problems because of higher natural moisture contents due to capillarity. This knowledge, plus data on seasonal high and low water tables, helps in deciding whether a granular layer is needed to provide a capillary break.

Site topographic surveys and grading plans establish surface drainage patterns. Along with historical data on surface flooding, this helps the designer to anticipate the worst possible moisture conditions at the site. It is also important to consider how ongoing development of the surrounding area will affect the future water table level and water movement on site.

For proper drainage, exterior finish grades should be at or below the floor subgrade level whenever possible (see Fig. 3-2). If the bottom of the granular subbase is below the adjacent finished grade, it could become a reservoir for water. The exterior grade also should slope away from the structure at a minimum slope of 40 mm per meter (1/2 in. per ft) for a distance of 2-1/2 m to 3 m (8 ft to 10 ft) beyond the building foundation.

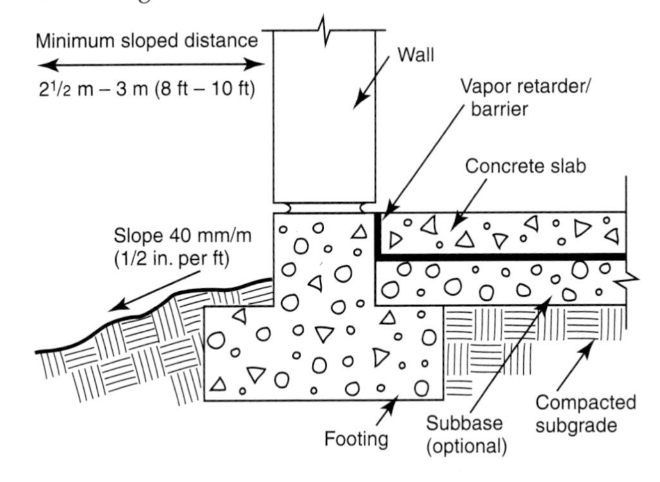

Fig. 3-2. Site grading to promote adequate drainage near buildings. Adapted from ASTM E 1643.

Subgrade and Subbase Design

Special design features are needed when the subgrade elevation is below the exterior finish grade. If the elevation difference is slight, the design detail shown in Fig. 3-3 is suggested. This detail includes a waterproofing membrane system on the outside of the wall and footing and an exterior drainpipe encased in granular fill. Drainage details for floors with subgrade levels substantially below grade are beyond the scope of this document.

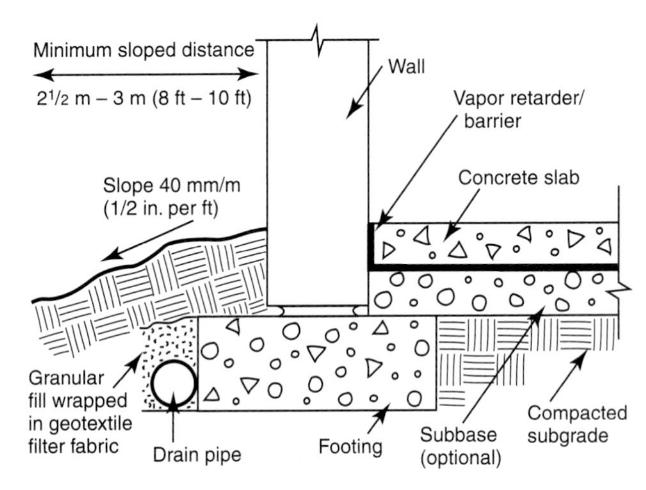

Fig. 3-3. Design detail when the subgrade is slightly below exterior finish grade. Adapted from ASTM E 1643.

A layer of open-graded, clean, compacted coarse aggregate is usually needed as a capillary break when the floor will be built on sandy soils or very fine soils (more than 45% of the particles by weight pass the 75-micron [No. 200] sieve). The capillary break usually consists of a 100-mm to 200-mm (4-in. to 8-in.) thick layer of 19-mm (3/4-in.) washed, single-size aggregate (see Fig. 3-4). The uniform aggregate size allows for air pockets between particles and reduces the upward movement of water. Gravel is preferred to crushed stone when a vapor retarder is specified because the smoother particles are less likely to puncture or tear the vapor retarder. Alternately, a geotextile fabric can protect the vapor retarder from the aggregate particles. If the water table is far enough below the surface, capillary water will not reach the underside of the slab and there is little need for a capillary break.

Fig. 3-4. Subbase materials are spread over the ground (subgrade), then leveled and compacted to provide a stable work surface that will not settle during or after concrete placement. (69423)

Vapor Retarder Design

Floors on ground that are to receive any form of floor covering should be built with a vapor retarder below the slab (see Fig. 3-5). A vapor retarder may also be needed in areas with a high water table or other source of subsurface moisture.

Where some or all of the structure is located below the water table, it may be necessary to use a vapor *barrier* (rather than a vapor retarder), a membrane that virtually stops moisture from passing through it. True vapor barriers are products such as rugged reinforced membranes that do not allow the passage of water vapor; they have water-vapor transmission ratings of 0.00 perms (per sq m per hr or per sq ft per hr) when tested in accordance with ASTM E 96. These are generally multiple ply products that may include polyethylene, reinforcing fibers or boards, metallic foils, paper and/or bituminous materials (see Fig. 3-5). Proper performance of vapor barriers requires sealing of all laps and strict adherence to the manufacturers' recommendations.

Fig. 3-5. Vapor retarders and vapor barriers can be made from many types of materials, but all serve the same purpose: to reduce or eliminate the passage of water vapor through concrete. (69683)

Vapor retarders effectively minimize water vapor transmission from the subbase through the slab, but are not 100% effective in preventing water vapor passage. Materials that have been used include 150-micrometer to 250-micrometer (6-mil to 10-mil) low-density polyethylene sheets and medium- to high-density polyethylene films. Some of these materials can be cross-laminated or reinforced. Because low-density polyethylene may deteriorate, it is recommended to use high-density polyethylene. Vapor retarders should meet the requirements of ASTM E 1745, *Standard Specification for Water Vapor Retarders Used in Contact with Soil or Granular Fill Under Concrete Slabs*. This standard sets requirements for water vapor permeance and establishes three grade classifications related to durability. Vapor retarders are required to have a permeance of less than 0.3 perms as determined by ASTM E 96.

Vapor retarders, especially thinner ones, are susceptible to damage. While a 1.25-mm (50-mil) sheet of polyethylene will most likely resist any type of construction traffic, a 100-micrometer (4-mil) sheet of polyethylene may be too thin to remain intact and may not provide long-term durability. During placement, thin sheets can be torn. During construction, they can be punctured when forms are placed, when steel is set, when concrete is placed, or by work traffic. As a result, both tensile strength and puncture resistance are important. (The practice of intentionally punching holes through a vapor retarder to allow bleedwater to exit the slab bottom is unacceptable because it defeats the purpose of the membrane.) A vapor retarder sheet placed directly on a crushed stone subbase is very likely to be damaged. Some specifiers recommend choking the top surface of the subbase with sand or crusher fines to prevent tearing or puncturing the vapor retarder. To further protect against punctures, some specifiers require a granular layer to be placed on top of the vapor retarder, between it and the concrete. This cushion also serves as a blotter layer and is discussed in more detail in the next section. Adequate lap and lap sealing at vapor retarder joints are also needed to ensure proper performance. ASTM E 1643, *Standard Practice for Installation of Water Vapor Retarders Used in Contact with Earth or Granular Fill Under Concrete Slabs,* covers surface preparation, placement, joint sealing, protection, and repair.

Vapor retarders installed for moisture control can also provide protection from radon gas. A limited discussion of radon, its effects on humans, and radon control in residential construction is given in Chapter 2, Subgrade Preparation.

Granular Layers over Vapor Barriers or Retarders

There is disagreement on the need for a granular layer between the vapor retarder and the concrete. Figure 3-6 provides a series of questions to consider when assessing whether or not a vapor retarder should be used. Figure 3-6 also helps a specifier decide where to place a vapor retarder (depending on why it is used). This decision should be made on a job-to-job basis.

Some specifiers always recommend the use of a vapor retarder under a slab. A vapor retarder prevents the subbase or subgrade from absorbing water from the fresh concrete and prolongs the period during which bleedwater accumulates at the surface. The longer bleeding period delays finishing or may cause related problems such as blistering or sheet scaling (delamination) if the concrete is finished too soon. Settlement cracking over reinforcing steel is another possible result of a longer bleeding period. One study linked increased plastic shrinkage cracking to concrete placed directly on a vapor retarder (Nicholson 1981). This study, however, evaluated concrete mixes with high water-cement ratios (greater than 0.7) representative of extreme conditions.

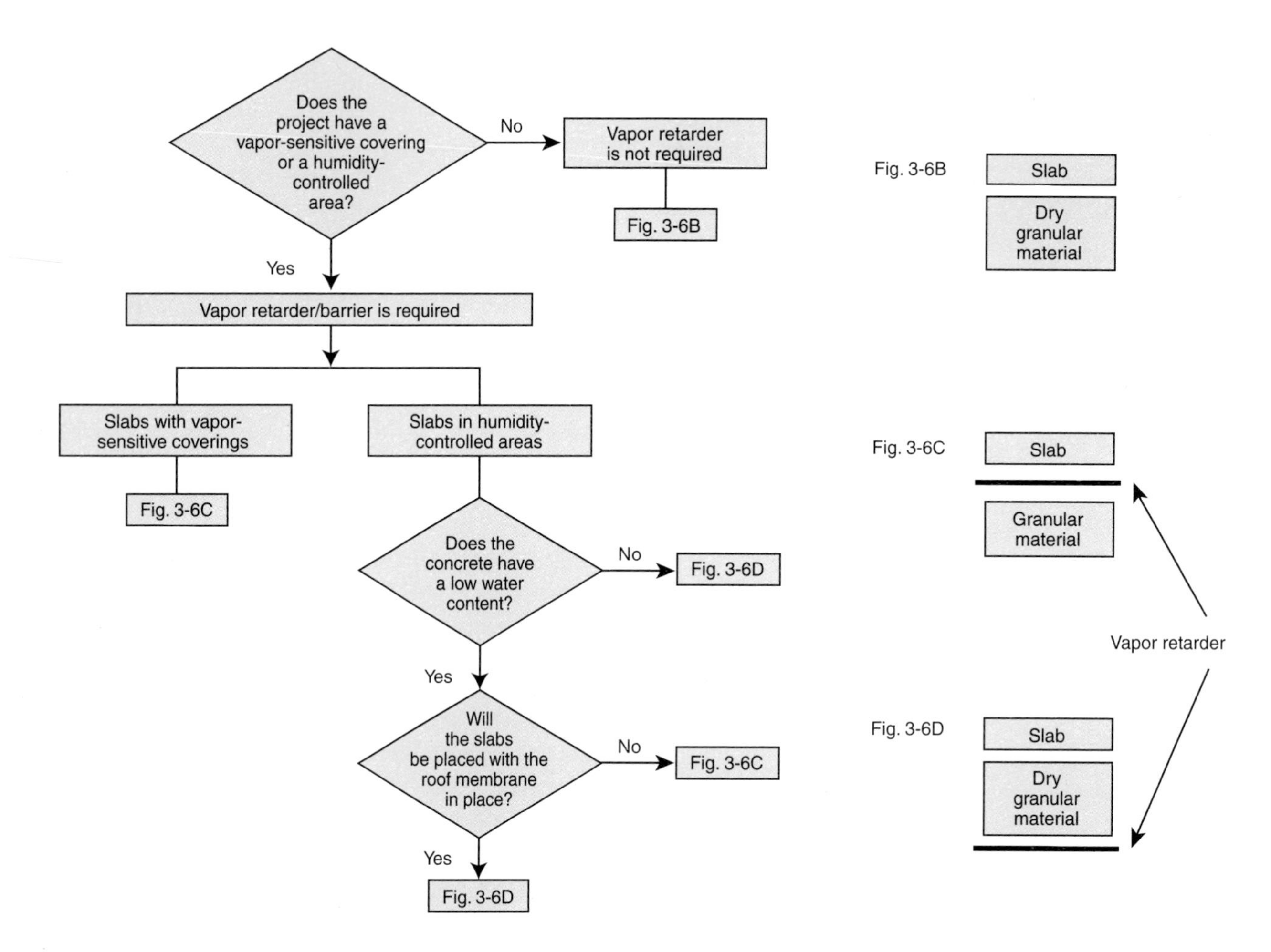

Fig. 3-6. A flow chart shows some steps to consider when deciding if a vapor retarder should be specified and where and how it should be placed. Adapted from Suprenant and Malisch 1998.

Placing concrete directly on the vapor retarder can also cause hardened concrete problems. Drying is not uniform in floors placed directly on vapor retarders. The top surface dries faster and thus shrinks more than the bottom. This can lead to curling of the slab edges and corners.

To minimize these problems, some specifications call for a blotter layer of granular material to be placed between the vapor retarder and the floor. A blotter layer can offer these advantages:

- puncture protection of the vapor retarder
- reduced finishing delays because the bleeding period is shorter
- less chance of blistering or sheet scaling
- less chance of settlement cracking over reinforcing steel
- less curling of the floor when it dries
- reduced likelihood of plastic shrinkage cracking
- possible increased concrete strength

To provide these benefits, it is recommended to use a 75-mm to 100-mm (3-in. to 4-in.) layer of trimmable, compactible, self-draining granular fill for the blotter layer. Concrete sand (ASTM C 33) is not recommended for this purpose because it does not provide a stable work platform. Concrete placement and workers walking on the sand can disturb the sand surface enough to cause irregular floor thickness and possibly create sand lenses in the concrete.

Other specifiers believe that no blotter layer is needed and that concrete should be placed directly on the vapor retarder. The idea is that concrete slabs should be cured from both the top and bottom. A granular layer between the vapor retarder and the concrete creates a potential water reservoir that could cause moisture problems in floors with coverings. Because more soil must be removed to accommodate the additional 75-mm to 100-mm (3-in. to 4-in.) blotter layer, the layer is more likely to be placed below finished grade level, increasing the chance of its holding water. Furthermore, an uncovered vapor retarder acts as a slip sheet that reduces slab restraint and thereby reduces random cracking.

Concrete placed directly on the vapor retarder can offer these advantages:

- reduced costs—less excavation, no need for additional granular material
- better curing of the slab bottom
- less chance of moisture problems with floor covering materials caused by water trapped in the granular layer
- low friction between the concrete and vapor retarder—reduced cracking

When placing concrete directly on a vapor retarder, it is recommended to use a low-water-content concrete and water-reducing admixtures to reduce bleeding, shrinkage, and curling of concrete.

CONSTRUCTION PRACTICES

When the design calls for a low-permeability covering to be placed on a concrete floor, special care may be needed during construction to control moisture content of the subgrade, subbase, or blotter layer (when used) over the vapor retarder. Wherever possible, the floor should be placed after the building is enclosed and watertight. This reduces floor moisture problems resulting from rainwater that might rewet the slab and saturate the subgrade, capillary break, or other granular layers.

Excessive sprinkling of a blotter layer before concrete placement can delay drying of the concrete floor. Wet curing methods such as ponding or continuous sprinkling will allow water to enter random slab cracks, or cracks below control or construction joints, and to pond in column blockouts, also contributing to a high moisture content beneath the floor slab.

Design details and construction methods affect drying rates of concrete floors and also affect performance of floors with low-permeability coverings. The steps described in this chapter are based on currently accepted practices for proper ground preparation and concrete placement and should help to control performance problems related to floor moisture.

CHAPTER 4

CONCRETE FOR FLOORS

The common function shared by most types of concrete floors is to provide support for applied loads, including people, vehicles, and goods. In addition, concrete often serves as the wearing surface. Good quality concrete is needed to carry loads and resist wear (see Fig. 4-1). The ingredients, the way they are combined, and the manner in which the concrete is placed all affect the quality and resulting performance of the floor. This section describes appropriate ingredients for concrete mixtures and fresh and hardened concrete properties. A detailed look at concrete performance allows drawing interrelationships between ingredients and properties.

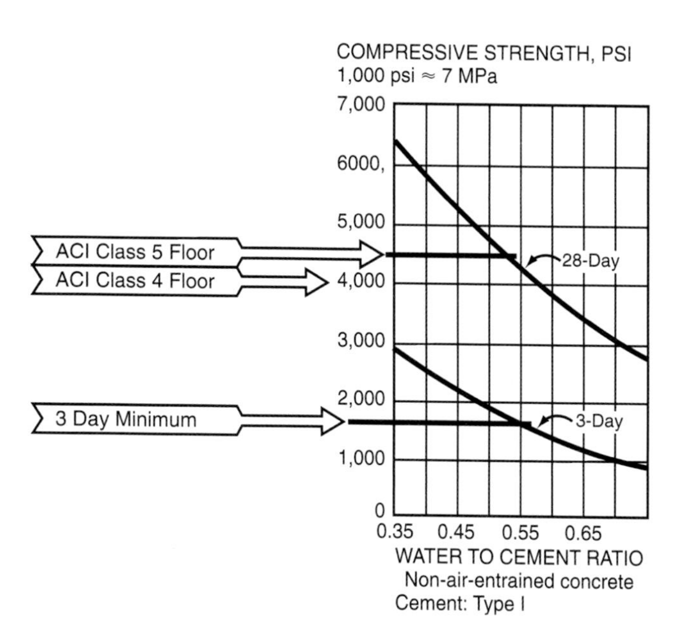

Fig. 4-1. ACI-recommended minimum concrete strengths for institutional/commercial (Class 4) and light-duty industrial (Class 5) floors on ground. (ACI 302)

INGREDIENTS AND CHARACTERISTICS

Cement Type and Content

Cement is the binder used to hold together all concrete ingredients. Cement specifications for portland and blended

Table 4-1. Applications for Commonly Used Cements

Cement specification	Applications*					
	General purpose	Moderate heat of hydration	High early strength	Moderate sulfate resistance	High sulfate resistance	Resistance to alkali-silica reactivity**
ASTM C 150 portland cements	I	II (moderate heat option)	III	II	V	Low alkali option
ASTM C 595 blended hydraulic cements	IS IP I(PM) I(SM) P	IS(MH) IP(MH) I(PM)(MH) I(SM()MH)		IS(MS) IP(MS) P(MS) I(PM)(MS) I(SM)(MS)		Low reactivity option
ASTM C 1157 hydraulic cements***	GU	MH	HE	MS	HS	Option R

* Check the local availability of specific cements as all cements are not available everywhere.
** The option for low reactivity with ASR susceptible aggregates can be applied to any cement type in the columns to the left.
*** For ASTM C 1157 cements, the nomenclature of hydraulic cement, portland cement, air-entraining portland cement, modified portland cement, or blended hydraulic cement is used with the type designation.

cements are found in ASTM C 150, ASTM C 595, and ASTM C 1157 (see Table 4-1).

If there are no special construction needs, a normal-use portland or blended cement is regularly used, as these are the most readily available. In ASTM C 150, this would be a Type I, II, or I/II; in C 595, Types IS, IP, P, I(PM), and I(SM); and in C 1157, Type GU. A single cement may simultaneously meet the requirements of more than one of these classifications.

Sometimes, construction needs or service environment dictate that a special cement type be used. Regarding construction needs, a fairly common requirement is putting the floor into service as quickly as possible. In this case, the use of high early strength cement may be warranted. Regarding service environment, aggressive exposures may suggest the use of a special type of cement for durability. For instance, high-sulfate environments can be due to soils in certain regions of the country or can be the result of certain industrial processes. Sulfate-resistant cements are available to meet the need of durable concrete in these environments.

When sulfate exposures are encountered in natural soils or in an industrial facility, cement should be chosen for its ability to resist sulfate attack. The severity of the exposure dictates the choice of cement that should be used to provide sulfate resistance to the concrete. Cement options for this purpose are shown in Table 4-2. Research has shown that low water-cementitious materials ratios are critical in the role of concrete durability in moderate or severe sulfate environments (see Fig. 4-2).

With modern concrete technology, higher strengths can now be obtained with less cement than in the past. Where strength alone is the decisive criterion, less cement means greater economy. Wear resistance of concrete, how-

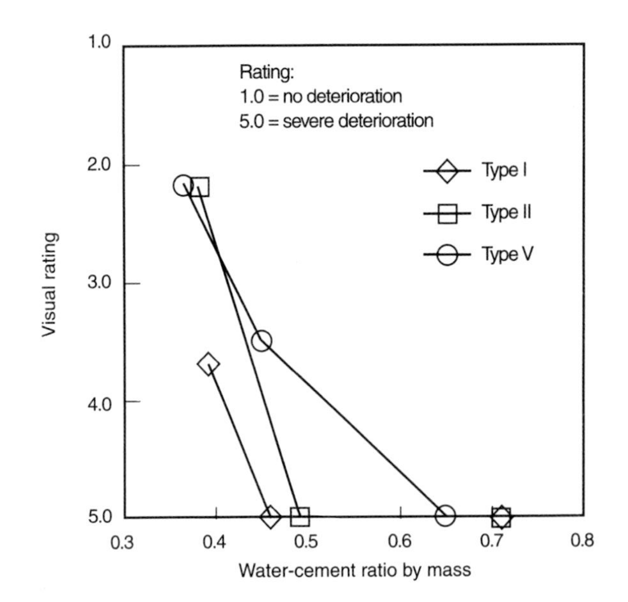

Fig. 4-2. (top) Both the cement type and the water-cement ratio affect concrete's ability to resist sulfate environments. (bottom) Beams in sulfate-soil test plot with durability illustrated in top graph. (66900)

Table 4-2. Requirements for Concrete Exposed to Sulfate-Containing Solutions*

Sulfate exposure	Water-soluble sulfate (SO_4) in soil, % by weight**	Sulfate (SO_4) in water, ppm**	Cement type***	Maximum water-cementitious materials ratio	Minimum compressive strength, MPa (psi)
Negligible	0.00–0.10	0–150	No restriction	—	—
Moderate†	0.10–0.20	150–1500	II, IP(MS), IS(MS), P(MS), I(PM)(MS), I(SM)(MS), MS	0.50	28 (4000)
Severe	0.20–2.00	1500–10,000	V, HS	0.45	31 (4500)
Very severe	Over 2.00	Over 10,000	V, HS	0.40	34 (5000)

* Adapted from References [Kosmatka and Panarese 1992, U.S. BUREC 1975, and ACI 318]

** Test procedure: *Method for Determining the Quantity of Soluble Sulfate in Solid (Soil or Rock) and Water Samples*, Bureau of Reclamation, 1977. ASTM is in the process of standardizing this method.

*** Cement Types II and V are specified in ASTM C 150, blended cements with the (MS)-suffix are specified in ASTM C 595, and the blended cements designated as MS or HS are specified in ASTM C 1157. Pozzolans (ASTM C 618), slag (ASTM C 989), or silica fume (ASTM C 1240) that have been determined by test or service record to improve sulfate resistance can also be used.

† Seawater.

Note: See (Kosmatka 1988) for alternatives when Type V or Type HS cement is not available. One option discussed is to use a cement and mineral admixture combination that has a maximum expansion of 0.10% at one year when tested according to ASTM C 1012, *Test Method for Length Change of Hydraulic-Cement Mortars Exposed to a Sulfate Solution*. Another option is to reduce the water to cementitious materials ratio to 0.35 or less.

ever, depends upon surface hardness as well as internal strength, and requires more cement than might be needed for strength alone. In order for finishing operations to progress smoothly, a floor surface requires sufficient cement paste. This is assured by specifying a minimum cement content. A minimum cement content also ensures adequate internal strength of the concrete.

The amount of cement should not be less than shown in Tables 4-3a and b. As indicated, using the largest possible aggregate size allows using lower cement contents.

Supplementary Cementing Materials

In addition to plain portland cement, other binders are available for use in concrete. Concrete mixtures for floors are regularly made with portland cement plus a supplementary cementing material (SCM), with blended cement, or with blended cement plus a supplementary cementing material. The common SCMs include fly ash, silica fume, calcined clay, and slag. Blended cements are formulated by combining one or more of these materials with portland cement, either before, during, or after grinding. SCMs are increasingly common ingredients in concrete, largely because of the benefits they provide:

- improved strength
- reduced permeability
- improved durability (such as reducing ASR)
- improved placing or finishing characteristics
- reduced cement content of the concrete

SCMs often improve the strength of concrete at later ages, but may slow the rate of strength gain at early ages, especially in cold weather (set times are extended except for silica fume). Concrete performance varies consider-ably with material characteristics and with combinations of materials. Chemical admixtures (retarders, accelerators) can be added to offset the differences incurred by the use of SCMs. Test mixes should be made to establish proportions, finishing behavior, setting characteristics, rate of strength gain, and ultimate strength. Availability of materials also influences which ones are chosen.

Pozzolans and slag, in addition to adding strength to concrete, help reduce the permeability and can increase durability in certain environments. Examples of potentially severe environments include sulfate exposures and abrasive environments.

Concrete finishers often appreciate the addition of fly ash or slag to a concrete mixture because it improves finishing characteristics. Silica fume, on the other hand, can make a mix sticky at high dosages. When using silica fume concrete mixtures, it may be necessary to modify the finishing operations. Floating should be accomplished immediately after strike-off, and since silica fume mixes do not bleed, fogging the surface may be necessary to prevent plastic shrinkage cracking.

In general, fly ash is used at a dosage of 15% to 30% of the total cementitious material. Silica fume typically accounts for between 5% and 10% of the total mass of cementitious material, while slag dosages range from 25% to 50%.

SCM usage can lead to more economical mixtures. For example, a concrete mixture containing a maximum aggregate size of 19 mm requires 320-kg cement per Table 4-3a. This concrete could be made with 320 kg of portland cement, or it could be made with 300 kg of portland or blended cement plus 20 kg of fly ash per cubic meter of concrete. Depending on the specifications, the fly ash dosage could be higher.

Table 4-3a. Minimum Cement Content (Metric)

Nominal maximum size of aggregate, mm	Cement,* kg/m^3
37.5	280
25.0	310
19.0	320
12.5	350
9.5	360

* Include supplementary cementing materials as part of the cement value. If SCMs are included in the cement content, the water-cement ratio (w/c) is more commonly referred to as water-cementitious materials ratio (w/cm).

Table 4-3b. Minimum Cement Content (Inch-Pound)

Nominal maximum size of aggregate, (in.)	Cement,* lb/yd^3
1-1/2	470
1	520
3/4	540
1/2	590
3/8	610

* Include supplementary cementing materials as part of the cement value. If SCMs are included in the cement content, the water-cement ratio (w/c) is more commonly referred to as water-cementitious materials ratio (w/cm).

Coarse Aggregate

Coarse aggregate shall meet the requirements of ASTM C 33. Floors generally perform better when random shrinkage cracking is minimal. The size and amount of coarse aggregate in the concrete influences random cracking. The larger the coarse aggregate particle, the better it restrains cement paste shrinkage. Also, for a given mass of material, the larger the coarse aggregate size, the lower the surface area of the aggregate. This permits using the lowest water content to mix the concrete, and low water content reduces random shrinkage cracking. Another factor that keeps water contents to a minimum is the use of well-graded aggregates—both coarse and fine—thus minimizing random shrinkage cracking.

Concrete has the lowest shrinkage potential when it contains:

- the largest possible size of coarse aggregate
- the largest amount of coarse aggregate in the mix proportions, while maintaining workability
- well graded aggregate to minimize voids between aggregate particles

Table 4-3 shows cement contents corresponding to aggregate size. The largest possible aggregate size should be used if it is economically available and if it satisfies placing and finishing requirements. The maximum aggregate size should not exceed:

- three-fourths the clear space between reinforcing bars
- one-third the depth of the slab

Fine Aggregate

Fine aggregate is composed of natural sand, manufactured sand, or a combination of the two. It fills spaces between coarse aggregate particles in concrete to make a more tightly packed mix. Along with cement and water, fine aggregate is part of the mortar that becomes the finished floor surface. Physical characteristics of the fine aggregate influence its performance in concrete. In a hard troweled floor, the very thin surface layer is paste (cement plus water) from the concrete mix. Just below it is the mortar layer (paste plus fine aggregate). Usually the concrete coarse aggregate is more depressed below the surface. If a floor is not hard troweled, the surface consists of mortar because paste is not brought to the surface.

Grading is one of the most important characteristics of fine aggregate because it affects water demand, workability, bleeding, and finishability. ASTM C 33 and C 330 provide standard gradations that can be used for most concrete mixtures. The C 33 limits shown in Table 4-4 are generally satisfactory for most concretes.

The fineness modulus (FM) is a common measure of gradation. FM is calculated by adding the cumulative percent material retained on the sieve sizes noted above, then dividing by 100. Sand should have an FM between 2.3 and 3.1 if it is to be used in concrete. A higher number indicates a coarser material. Once a material is chosen, its fineness modulus should not vary by more than 0.2 during floor placement. For hand-finished floors or where a smooth surface texture is desired, fine aggregate with at least 15% passing the 300 µm (No. 50) sieve and 3% or more passing the 150 µm (No. 100) sieve should be used. If finer cements, SCMs, or moderate to high cement contents are used, it may be possible (and even desirable) to reduce the amount of material passing these two sieves with no noticeable difference in quality of finish.

Table 4-4. Fine Aggregate Grading (ASTM C 33)

Sieve size		Percent passing by mass
9.5 mm	3/8 in.	100
4.75 mm	No. 4	95 to 100
2.36 mm	No. 8	80 to 100
1.18 mm	No. 16	50 to 85
600 µm	No. 30	25 to 60
300 µm	No. 50	10 to 30
150 µm	No. 100	2 to 10

Particle shape may also affect finishability: rounded particles are generally better for finishing, whereas flat or elongated particles may make finishing difficult (Garber 1991). However, there is no way to predict the aggregate's behavior, so the best indicator of performance in concrete is experience with the specific materials.

Chemical Admixtures

Good quality, basic concrete, of nominal strength and durability, can be made without admixtures. However, chemical admixtures offer a number of advantages in concrete construction that include improving concrete properties and construction procedures:

- reduced bleeding
- improved setting and finishing characteristics
- less cracking
- increased strength, durability
- uninterrupted hot or cold weather construction
- economical mixes

Since admixtures are best used to enhance a mix, a suitable admixture should be identified only after choosing cementing materials and aggregates that meet the needs of the job.

Chemical admixtures in North America should conform to ASTM C 494 and C 1017. Compliance with ASTM, however, does not guarantee that the admixture will achieve the desired result. Consult the manufacturer's representatives and product literature before mixing test batches to verify performance. Materials include water-reducing, retarding, accelerating, and superplasticizing admixtures, or combinations thereof. Air-entraining admixtures have their own specification, ASTM C 260. Chemical admixtures may reduce or increase concrete shrinkage. Shrinkage reducing admixtures are a newer class of admixtures whose ultimate purpose is to reduce cracking. No ASTM designation yet exists for shrinkage reducing admixtures.

Water-reducing admixtures are usually specified for one of two purposes: maintaining workability or decreasing the water-cementitious materials ratio. Designers often have to limit the water-cementitious materials ratio (w/cm) to improve both strength and durability. At low water-cementitious materials ratios, a mix can be hard to place and consolidate, but the addition of a water-reducing admixture can change that and provide the needed workability. Another use of a water-reducing admixture is to remove a portion of the cementing materials while maintaining strength and workability, thus increasing the economy of the mix. This can only be done if the reduction in cement content is acceptable from the standpoint of floor finish and durability.

Retarding admixtures are beneficial in hot weather, when temperatures are warm enough to speed up the initial set. If set time decreases, there may not be enough time available for finishing. A retarder can delay the set of the concrete to allow construction to proceed at a normal pace.

Water-reducing admixtures and retarding admixtures often are based on similar chemicals. When using these chemicals, it should be recognized that the other effect may be introduced to the concrete. This may be desirable or undesirable. To call attention to the distinction, some water-reducing admixtures are designated by a secondary function: water-reducing/retarding or water-reducing/accelerating.

Accelerating admixtures have the opposite effect of retarders and are used in cold weather. Because cold temperatures slow the set of concrete, it may take a long time to reach initial set, and this can place unreasonable demands on the finishing crew. An accelerator helps combat the problem.

Air-entraining admixtures are used primarily for freeze-thaw protection. A secondary use is improved workability of fresh concrete. Whereas air-entraining admixtures have an excellent track record in outdoor slabs, these slabs often receive a minimum of finishing. The admixtures should be used with caution for interior floors on ground, only being specified when they are absolutely needed. Interior slabs often receive a great deal of finishing, and the addition of air entrainment may lead to finishing difficulties or problems with delaminations, blisters, or other surface defects.

Commercial shrinkage-reducing admixtures were introduced in the 1990s. Manufacturers claim they typically provide a 25% to 50% reduction in the ultimate shrinkage of concrete. Less shrinkage should lead to a reduction in the number and/or size of cracks. This could be beneficial in certain applications where cracks due to drying shrinkage should be minimized, such as on industrial floors. There are many approaches for reducing drying shrinkage, such as keeping the water content to a minimum and maximizing the coarse aggregate size and content.

Fibers

Many types of fibers are used in concrete (PCA 1991). Common fibers for floor applications are made of plastic. Steel fibers, providing high flexural strengths and impact resistance, are found in heavy-duty industrial floors and other floors.

Fibers can improve concrete's flexural strength, ductility, toughness, impact resistance, fatigue, and wear resistance. Depending on the fiber type and dosage, fibers can reduce plastic shrinkage cracking in fresh concrete.

Fibers are identified according to their length, diameter, specific gravity, tensile strength, and modulus of elasticity. Fibers are generally added to fresh concrete by the ready-mix producer and dispersed by the action of mixing. *In no case should extra water be added to compensate for apparent reduced workability caused by the addition of fibers.* Chemical admixtures or other mix adjustments should be considered to improve fresh handling properties. Joints in the concrete must be handled in the same way as if the fibers were not present.

Plastic fibers are generally dosed at 0.1% to 0.3% by volume for low-volume percentages and 0.4% to 0.8% for high-volume percentages. The 0.1%-level, common for slabs on grade, is equivalent to 0.9 kg/m^3 (1.5 lb per yd^3). At this dosage, physical properties of the concrete can be marginally improved, and plastic shrinkage cracking (if it is a problem) may be reduced (ACI 544.1R). High-volume percentages can provide substantial increases in ductility and impact resistance. Plastic fibers should conform to the requirements of ASTM C 1116.

Steel fiber contents generally range from 0.5% to 2.0% by volume. Beyond 2% dosage, dispersion of the fibers in the mix is generally too poor and workability is greatly reduced. The properties of concrete most positively influenced by adding steel fibers are flexural strength and toughness (PCA 1991). Steel fibers should conform to the requirements of ASTM A 820.

Concrete Tolerances

Concrete proportioning, slump, and strength tolerances should be as follows:

- cement by weight ± 1%
- added water by weight ± 1%
- fine and coarse aggregate by weight ± 2%
- admixtures by weight ± 3%
- slump at stated maximum or up to 25 mm (1 in.) less

Strength is acceptable if the average strength of all sets of three consecutive strength tests exceed the specified strength and no individual result is less than 3.4 MPa (500 PSI) of specified strength.

FRESH AND HARDENED PROPERTIES

Fresh concrete properties affect both how the concrete is handled and the hardened concrete characteristics. For floors, hardened concrete should be able to carry loads and resist wear. Both load-carrying capacity and wear resistance are affected by compressive strength of the concrete.

Slump

Excessive water used to produce high slump is a primary cause of poor floor performance, as it leads to bleeding, segregation, and increased drying shrinkage. If a finished floor is to be level, uniform in appearance, and wear resistant, all batches placed in the floor must have nearly the same slump and must meet specification criteria.

Low-slump (50 mm to 100 mm [2 in. to 4 in.]) concrete flatwork is routinely struck off with mechanical equipment like vibratory screeds. Typically, using such equipment for floor work facilitates concrete consolidation, requiring less water to be added at the jobsite, which ultimately results in improved wear resistance of the surface. Low-slump concrete (see Fig. 4-3) helps to:

- reduce finishing time
- reduce cracking
- minimize surface defects

Recommended strengths and slumps for each ACI class of floor are given in Table 4-5, adapted from ACI 302.1R. This slump limit is the same whether or not the concrete contains chemical admixtures (superplasticizers or normal water-reducing admixtures). Table 1-1 defines all nine floor classes.

Fig. 4-3. Low- to moderate-slump concrete alleviates many concerns associated with concrete floor placement and finishing. (44485)

Air Content

Concrete for floors is usually not air entrained. A small amount of entrained air is sometimes useful for concrete floors because it reduces bleeding and increases plasticity. A total air content (including both entrapped and entrained air) of 2% to 3% is suggested. Concrete that will be exposed to cycles of freezing and thawing and the application of deicer chemicals requires a greater total air content—about 5% to 8% depending upon maximum size of aggregate—to ensure resistance to scaling. For most floors discussed in this publication, the maximum size aggregate used will be between 9.5 mm and 37.5 mm [3/8 in. and 1-

Table 4-5. Recommended Strength and Slump for Floors

Floor class*	Surface traffic	Slab surface	28-day compressive strength		Maximum slump**	
			MPa	psi	mm	in.
1	foot	exposed	21.0	3000	125	5
2		covered	21.0	3000	125	5
3		exposed	28.0	4000	125	5
3		covered	21.0	3000	125	5
4	light vehicle	exposed	28.0	4000	125	5
4		covered	24.5	3500	125	5
5	soft solid wheels	exposed	31.5§§	4500§§	75	3
6	hard wheels	not†	35.0§§	5000§§	75	3
6		hardener†	31.5	4500	75	3
7		base	28.0§§	4000§§	125	5
7		topping	35.0+	5000+	75	3
8	as for Class 4-6	base	28.0+§§	4000+§§	125	5
8		topping††, §	35.0+	5000+	75	3
9		superflat††	31.5§§	4500§§	75	3

* Refer to Table 1-1 for floor class definitions.
** Target slumps should be 25 mm (1 in.) less than the maximum shown to allow for mix variation. Some floor classes allow up to 125-mm (5-in.) slump, but lower slumps will result in less shrinkage, less curling, and other desirable properties.
† Metallic or mineral aggregate surface hardener.
†† The strength required will depend on the severity of usage.
§ Maximum aggregate size not greater than one-quarter the thickness of unbonded topping.
§§ Minimum 28-day flexural strength of 4 MPa (600 psi) (solid and hard wheels include heavy vehicle loading).

1/2 in.]. However, it should be recognized that floating and troweling, depending on finishing intensity, may remove a significant amount (or all) of the entrained air from the concrete mortar at the surface, severely impairing freeze-thaw durability.

Whereas minimum air contents are well established for durability (resistance to freeze-thaw or deicer attack), there is a reason to consider setting a maximum: when floor finishing operations include steel troweling. A maximum total air content of 3% has been established to reduce the possibility of blistering. This occurs because steel trowels can seal the surface and trap air pockets beneath it, especially when monolithic surface treatments are used.

Air content is chosen based on the needs of construction (ease of placement) and service environment (exposure). The air content of the fresh concrete must be tested if compliance with specifications is mandatory. Though testing is recommended, it is not always performed on smaller projects. *Regardless of project specifications,* if the floor will be subjected to traffic, total air content should be measured at the point of placement to verify compliance with the specified air content. Unintentionally entrained air might push the total air content above an acceptable level. Dry-shake surfaces, especially, are at risk of blistering if the total air content exceeds 3%.

Testing can be done using a pressure air meter (see Fig. 4-4) or a volumetric meter. Building codes and standards indicate air content testing frequency, which is at least one test daily. Only a qualified inspector should run an air test for verifying conformance to specifications.

Whenever entrained air is used in concrete that will receive a steel trowel finish, precautions must be taken to prevent surface blistering. Air-entrained concrete bleeds more slowly than non-air-entrained concrete, so the surface may appear ready for floating and troweling while the underlying concrete is still bleeding or plastic and releasing air. If finishers seal the surface by troweling before bleeding has ceased, blisters can form below a dense troweled skin of mortar about 3 mm (1/8 in.) thick. Blisters form late in the finishing process, after floating and the first troweling.

To avoid blistering, place, strike off, and float the concrete as rapidly as possible without working up an excessive layer of cement paste. Keep the float blades flat in initial floating to avoid densifying the surface too early. After these operations are completed, delay further finishing as long as possible by covering the surface with polyethylene or otherwise protecting it from evaporation.

If these measures do not prevent blistering, avoid using an air-entraining admixture in the concrete.

Do not use air-entrained concrete for floors that will receive an application of dry-shake surface hardener. These products require some moisture at the slab surface so they can be thoroughly worked in. Because the entrained air slows bleeding, needed moisture may not be present, and blistering or delamination of the hardened surface is more likely.

Fig. 4-4. A pressure meter is the most common way of checking the air content of a fresh concrete mixture. (66113)

Bleeding

In concrete construction, bleeding is the development of a layer of surface water caused by the settling of solid particles (cement and aggregate) and the simultaneous upward migration of water (Fig. 4-5). Some bleeding is normal and helps to control plastic shrinkage cracking, but excessive bleeding increases the water-cement ratio near the surface, particularly if finishing operations take place while bleedwater is present. This may result in a weak surface with poor durability (Kosmatka 1994).

The amount and rate of bleeding increases with higher initial water content and thicker floors. The following means can be used to reduce bleeding:

- properly graded aggregate
- supplementary cementitious materials
- finer cements
- certain chemical admixtures
- air entrainment

Shrinkage

Cracking can be the result of one or a combination of factors such as drying shrinkage, thermal contraction,

Fig. 4-5. Bleeding brings water to the concrete surface. (P29992)

restraint (external or internal) to shortening, subgrade settlement, and applied loads. Installing joints in concrete floors forces cracks to occur in places where they are inconspicuous and thus controls random cracking.

Cracks that occur before hardening are usually the result of settlement within the concrete mass, or shrinkage of the surface caused by rapid loss of water while the concrete is still plastic. These are referred to as plastic-shrinkage cracks.

As the concrete settles or subsides, cracks may develop over embedded items, such as reinforcing steel, or adjacent to edges where it contacts forms or hardened concrete. These cracks are referred to as settlement cracks. Settlement cracking results from insufficient consolidation (vibration), high slumps (overly wet concrete), or a lack of adequate cover over embedded items.

Plastic-shrinkage cracks are relatively short cracks that may occur before final finishing on days when one or more of the following conditions exist; wind, low humidity, and high temperature. Under these conditions, surface moisture evaporates faster than it can be replaced by rising bleedwater. As a result, the surface of the concrete sets before lower portions of the slab. As the surface hardens it begins to shrink more than the concrete below, allowing plastic shrinkage cracks to develop in the slab surface. Plastic shrinkage cracks often penetrate to mid-depth of a slab. They vary in length and are usually parallel to each other, spaced from a few centimeters (inches) up to 3 m (10 ft) apart.

Cracks that occur after hardening are usually the result of drying shrinkage, thermal contraction, or subgrade settlement. After hardening, concrete begins to dry and shrink as a result of water leaving the system. For small, unrestrained concrete specimens (cylinders), shrinkage (strain) has been measured at 500 to 800 millionths (at 50% relative humidity and 23°C [73°F])(Hanson 1968). "Real" concrete shrinks less. A floor (which is retrained by the subgrade) typically will be at 80% to 95% relative humidity so the shrinkage is actually closer to 100 to 300 millionths.

To accommodate shrinkage and control the location of cracks, joints are placed at regular intervals. Experience has shown that contraction joints (induced cracks) should be spaced at a distance from about 24 to 30 times the slab thickness. This is equivalent to 5-m to 6 -m (17-ft to 20-ft) intervals in each direction in 200-mm (8-in.) thick unreinforced concrete slabs on grade. If reinforcement is added, and if intermediate random joints are acceptable, longer joint spacings can be used. See Tables 6-1a and b for other slab thicknesses and corresponding joint spacings.

The major factor influencing the drying shrinkage of concrete is the total water content. As the concrete's water content increases, the amount of shrinkage increases proportionally. Large increases in the sand content and significant reductions in the size of the coarse aggregate increase shrinkage because total water is increased and because smaller size coarse aggregates provide less internal restraint to shrinkage. Use of high-shrinkage aggregates and calcium chloride admixtures also increase shrinkage. Some chemical admixtures complying with ASTM C 494 can increase shrinkage. Within the range of practical concrete mixes—280 to 445 kg/m^3 cement content (470 to 750 lb/yd^3, formerly referred to as "5- to 8-bag" mixes)—increases in cement content have little to no effect on shrinkage as long as the water requirement is not increased significantly.

Silica fume can make highly cohesive, sticky concrete, with little bleeding capacity. With little or no bleedwater on the surface, silica fume concrete is prone to plastic shrinkage cracking on hot, windy days. Fogging the air above the concrete and erecting windshades reduce the risk of plastic shrinkage cracking.

Concrete has a coefficient of thermal expansion and contraction of about 0.0000010 mm/mm/°C (0.0000055 in. per in. per °F). Concrete placed during hot midday temperatures will contract as it cools during the night. A 22°C (40°F) drop in temperature between day and night—not uncommon in some areas—would cause about 0.8 mm (1/32 in.) of contraction in a 3-m (10-ft) length of concrete, sufficient to cause cracking if the concrete is restrained. Thermal expansion can also cause cracking.

Cracking in concrete slabs on grade of correct thickness for the intended use can be reduced significantly or eliminated by observing the following practices:

- Use proper subgrade preparation, including uniform support and proper subbase material with adequate moisture content.
- Minimize the mix water content by maximizing the size and amount of coarse aggregate and by using low-shrinkage aggregate.
- Use the lowest amount of mix water required for workability; do not permit overly wet consistencies.
- Avoid calcium chloride admixtures.

- Prevent rapid loss of surface moisture while the concrete is still plastic through use of spray-applied finishing aids or plastic sheets to avoid plastic-shrinkage cracks.
- Provide contraction joints at reasonable intervals, 24 to 30 times the slab thickness.
- Provide isolation joints to prevent restraint from adjoining elements of a structure.
- Prevent extreme changes in temperature.
- To minimize cracking on top of vapor retarders, use a 100-mm (4-in.) thick layer of slightly damp, compactible, drainable fill choked off with fine-grade material. If concrete must be placed directly on polyethylene sheet or other vapor retarder, use a mix with a low water content.
- Properly place, consolidate, finish, and cure the concrete.
- Consider using a shrinkage reducing admixture to reduce drying shrinkage, which may reduce shrinkage cracking.
- Consider using plastic fibers to control plastic shrinkage cracking.

Proper mix design and selection of suitable concrete materials can significantly reduce or eliminate the formation of cracks.

Compressive Strength

For offices, residential, and commercial applications (classes 1-3), a minimum strength of 21 MPa (3000 psi) is required at 28 days (see Table 4-4). Lower strengths (less than 20 MPa) have been used successfully in applications not needing high degrees of wear resistance. A minimum strength of 28 MPa (4000 psi) at 28 days is advisable for most light-duty industrial and institutional/commercial floor use. For industrial floors subject to heavy traffic, a 28-day strength of 31 MPa (4500 psi) or higher is recommended. Lower strength may be adequate for supporting the loads on the floor, but can be inadequate for satisfactory wear resistance in highly abrasive environments. It is also advisable to require 12.5 MPa (1800 psi) strength before allowing light-weight construction traffic on the slab. Note: Concrete strength is affected by many factors. The curves shown in Fig. 4-6 represent typical strength values for concrete made with an ASTM C 150 Type I cement. Actual strengths on a job will vary from strength shown in the figure.

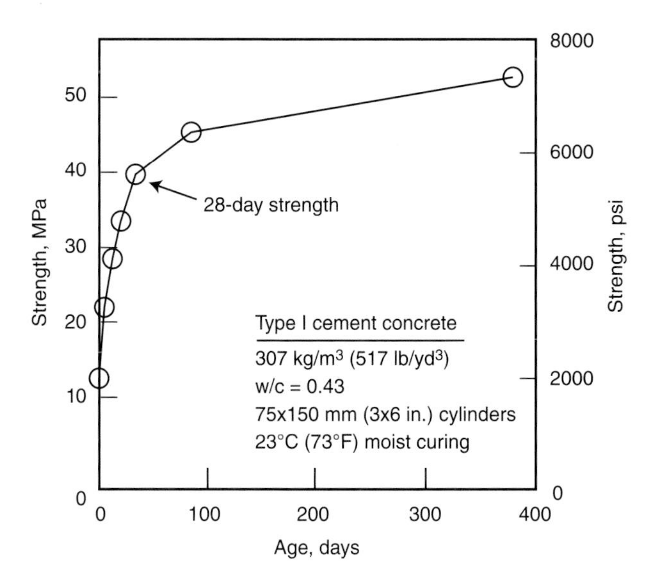

Fig. 4-6. Typical strength curve for concrete made with Type I cement. Data are shown for 1, 3, 7, 14, 28, and 91 days and for 1 year (Gebler and Klieger 1985).

Wear resistance of concrete floor surfaces, as it relates to rolling traffic from equipment with solid rubber tires, hard polyurethane wheels, and steel casters, is enhanced by hard troweling densification during floor finishing. Wear resistance is most important in the top 3 mm (1/8 in.) of the floor, and is very much influenced by the construction methods employed, especially finishing methods. Further discussion of finishing to provide hard, wear-resistant, and non-dusting surfaces is presented in Chapter 8 under "Burnished Floors."

Required strength should be no more than is necessary to meet the strength and wear resistance requirements of the slab. However, strengths greater than the minimums indicated above may permit a reduction in slab thickness and provide improved wear resistance.

Classification of floors on the basis of intended use is given in Table 1-1.

Flexural Strength

When a load is applied to a floor on ground, it produces stresses in the concrete slab (see Fig. 4-7). Compressive stresses caused by the load on the slab are typically much lower than the compressive strength. More important than the compressive stress is the flexural stress caused by loading. Flexure is critical because it places a portion of the concrete in tension, and the tensile strength of concrete is a small fraction of the compressive strength. Most slabs that

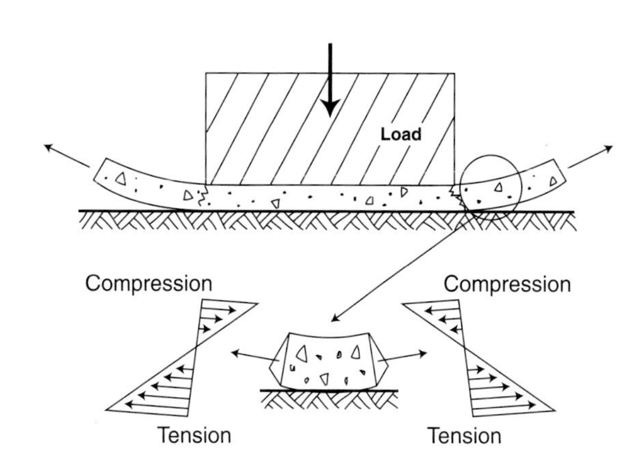

Fig. 4-7. Loads on a floor induce flexural stresses in a concrete slab.

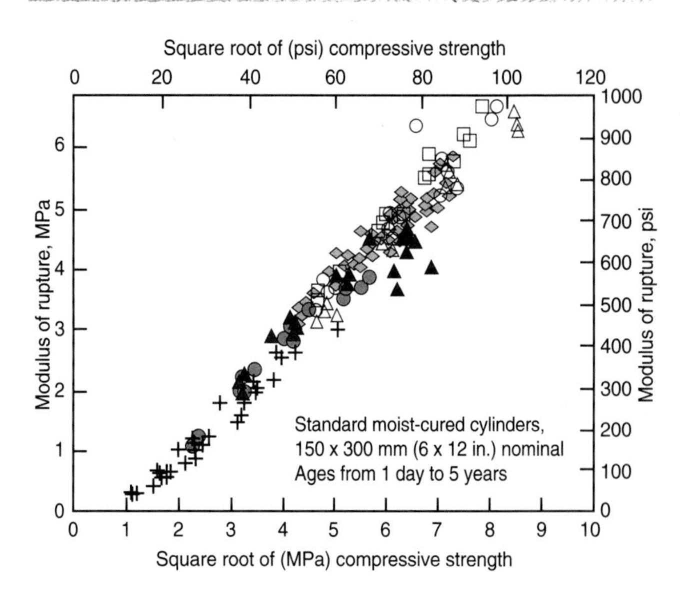

Fig. 4-8. (top) Third-point loading tests measure the flexural strength of concrete beam specimens in the lab or in the field with a portable flexural strength tester (shown here). Results from compressive strength tests of cylinders are correlated with flexural strength, useful in the quality control of concrete for floors. (middle) Companion beam and cylinder samples can be made and tested to deter-mine the relationship between flexural and compressive strengths. (bottom) Long-term data show that compressive strength is proportional to the square root of flexural strength (measured by third-point loading) over a wide range of strength levels (Wood 1992). (69684, 69653)

fail do so as a result of failure in flexure. Consequently, *the flexural stress and the flexural strength (modulus of rupture) of the concrete are used in floor design to determine slab thickness.*

Flexural strength is determined by modulus of rupture (MOR or MR) tests in accordance with ASTM C 78, *Flexural Strength of Concrete (Using Simple Beam with Third-Point Loading)* (see Fig. 4-8). Usually the 28-day strength is selected as the design strength for floors. This is generally conservative since the concrete continues to gain strength after 28 days. It may not be conservative when early-age construction loading occurs.

There are some difficulties associated with flexural strength testing. The test specimens are large, so they are somewhat difficult to handle. The measurement of flexural strength is quite sensitive to variations in test specimens and procedures, more so than compressive strength testing. Compressive strength cylinders (see Fig. 4-8) are easier to make and move, and are less prone to damage than flexural strength specimens. This is an important consideration for specimens made at the jobsite that will have to be transported to the lab for testing. For these reasons, a relationship is developed between flexural strength and compressive strength by laboratory testing. Compressive strength test results can then be used to estimate the flexural strength by the formula:

$$MR = k\sqrt{f_c}$$

MR = modulus of rupture or flexural strength, in MPa or psi

f_c = compressive strength, in MPa or psi

k (metric) = constant, usually between 0.7 (for rounded aggregate) and 0.8 (for crushed aggregate)

k (in.-lb) = constant, usually 9 to 11

Approximate correlations between compressive and flexural concrete strength (for the case where k = approx. 0.74, metric [k = 9 in in.-lb units]) are listed in Table 4-6a, b.

Evidence of the relationship between compressive strength and square root of the flexural strength is seen graphically in Fig. 4-8, which shows long-term data for a number of concrete mixtures.

Table 4-6a. Approximate Relationship Between Compressive and Flexural Strength of Concrete (Metric)

Compressive strength, f_c, MPa	Flexural strength, *MR*, MPa
28	3.9
31	4.1
34	4.3
38	4.6
41	4.7
45	5.0
48	5.1

Table 4-6b. Approximate Relationship Between Compressive and Flexural Strength of Concrete (In.-Lb Units)

Compressive strength, f_c, psi	Flexural strength, *MR*, psi
4,000	570
4,500	600
5,000	640
5,500	670
6,000	700
6,500	730
7,000	750

DURABILITY

There may be other considerations for concrete durability, depending on the service environment. While air entrainment is beneficial for concrete exposed to cycles of freeze-thaw, it must be used with caution (if at all) for interior concrete floors, especially ones that will receive coverings or special toppings. See ACI 302.1R for further information.

Wear Resistance

Research has shown that a reduced water-cement ratio *at the surface* improves wear resistance and strength. Mortar richness is important because coarse aggregate hardness and toughness become significant only after the surface-mortar matrix has worn away. For flatwork, concrete placeability and finishability are as important as strength because they have a significant effect on the quality of the top 3.0 mm (1/8 in.) of the wearing surface. Thus, strength alone is not a good measure of the concrete quality needed for flatwork.

To get the desired concrete quality, the ready-mixed concrete order must clearly state the basic information contained in job specifications. Strength, minimum cement content, coarse aggregate maximum size, and slump should all be specified. Air content should also be specified when applicable.

For outdoor applications where free-thaw occurs, there is no question that concrete should contain entrained air. The added air protects exposed flatwork from cycles of freezing and thawing and from deicing agents.

For interior applications where freeze-thaw occurs, the needs are different. A small amount of entrained air (about 3% total air content) is sometimes specified for interior floors to improve finishability by increasing plasticity and reducing bleeding and segregation. For a steel-trowel finish, most specifiers and floor constructors prefer to leave out the air-entraining admixture to reduce the risk of blistering and delaminations.

Chemical Resistance

Concrete floors can be exposed to many types of chemicals that can attack and deteriorate concrete. For instance, cleaning agents and substances that are used in chemical or meat processing plants may be harmful to concrete.

Good quality concrete is the first line of defense against chemical attack. All aspects of making concrete—ingredients, mixing, and handling—can have an affect on concrete's chemical resistance. While applying coatings or coverings in severe environments prevents corrosive substances from contacting the concrete, protective surface treatments are not infallible, as they can be damaged during construction or can deteriorate with use (traffic over the floor).

In general, two requirements for durable concrete are adequate strength and sufficiently low permeability.

For further information on the chemical resistance of concrete, see *Effects of Substances on Concrete and Guide to Protective Treatments* (PCA 2001).

Alkali-Aggregate Reaction

Aggregates containing certain constituents can react with alkali hydroxides in concrete. The reactivity is potentially harmful only when it produces significant expansion. Alkali-aggregate reactivity (AAR) has two forms—alkali-silica reaction (ASR), which is common, and alkali-carbonate reaction (ACR), which is rare. *Distress* resulting from ASR in structural concrete, however, is not common.

In floors, typical indicators of ASR presence might be popouts or a network of cracks. ASR popouts induced by sand particles can occur soon after a floor is placed. See Chapter 10 for more information.

In general, the alkali-silica reaction forms a gel that swells as it draws water from the surrounding cement paste. Reaction products from ASR have a great affinity for moisture. In absorbing water, these gels can induce pressure, expansion, and cracking of the aggregate and surrounding paste. The reaction can be visualized as a two-step process:

1. Alkali + reactive silica → gel reaction product (alkali-silica gel)
2. Gel reaction product + moisture → expansion

The amount of gel formed in the concrete depends on the amount and type of silica and alkali hydroxide concentration. The presence of gel does not always coincide with distress, and thus, gel presence doesn't necessarily indicate destructive ASR.

For alkali-silica reaction to occur, three conditions must be present:

- reactive forms of silica in the aggregate
- high-alkali (pH) pore solution
- sufficient moisture

The best way to avoid ASR is to take appropriate precautions before concrete is placed. Standard concrete specifications may require modification to address ASR. The most effective way of controlling expansion is to design mixes specifically to control ASR, preferably using locally available materials. In North America, current practices include using a pozzolan, slag, or blended cement proven by testing to control ASR or limiting the alkali content of the concrete.

See *Diagnosis and Control of Alkali-Aggregate Reactions in Concrete* (Farny and Kosmatka 1997) for more information.

In addition, it is prudent to check concrete mixtures for susceptibility to alkali-silica reaction (potentially reactive combinations of cement and aggregate). See PCA's IS413 and IS415 for guidance.

Fire Resistance

Hardened concrete can be exposed to elevated temperatures for extended periods without experiencing any deterioration. Usually, exposure temperatures of about 200°C (400°F) or less cause no damage to concrete. When the temperature goes above this limit, moisture begins to be driven out of the concrete, potentially resulting in strength loss, spalling, and other irreversible damage (Hanley-Wood 1971). At higher temperatures, the cement paste starts to dehydrate and aggregates begin to break down. The temperature at which aggregate deterioration commences is dependent on aggregate type.

Concrete fire resistance for floors (and walls and roofs) is measured in accordance with ASTM E 119, *Test Methods for Fire Tests of Building Construction and Materials.* The fire rating is the minimum length of time (in hours) that concrete will:

- ensure structural adequacy and protect human life
- prevent the spread of high temperatures between adjoining rooms/floors

Carbonate and lightweight aggregates usually perform slightly better than siliceous aggregates. Concrete made with either carbonate or lightweight aggregates can retain more than 75% of its original compressive strength at temperatures up to 650°C (1200°F). The corresponding temperature for the siliceous aggregate concrete is about 430°C (800°F) (Abrams 1973). As temperatures increase, strengths decrease. The percent of strength retained is independent of the original concrete strength.

There are a number of factors that affect the measured fire rating of concrete. Its moisture condition has a large influence on behavior at high temperatures. New concrete contains much moisture and should not be exposed to elevated temperatures until it has had a chance to dry. Otherwise, free water in the concrete pores can turn to vapor, an expansive process that can result in concrete spalls.

Concrete for high-temperature applications, called refractory concrete, should be specially designed. This includes the choice of aggregate and cement type. As noted above, some aggregates resist fire better than others. In general, carbonate aggregate performs better than siliceous aggregate, and lightweight or expanded aggregate performs better than normal-weight aggregate. Guidelines for choosing aggregates based on service temperature are found in ACI 547R, *Refractory Concrete.*

The size of the concrete member also has an effect on the structural behavior of the element. Independent of aggregate type, a 2-hour fire endurance is attainable if the floor thickness is 125 mm (5 in.) or greater (Abrams and Gustaferro 1968). Since two hours is a typical fire rating required by many building codes, and since many industrial and commercial floors often exceed this thickness for structural purposes, fire resistance is generally not the critical design criterion.

CHAPTER 5

FLOOR THICKNESS DESIGN

Many variables directly or indirectly influence the determination of thickness requirements for concrete floors on ground. To include all of them in a design method would be an unduly complex procedure and could lead to overconfidence in the design as a guarantee of good floor performance. Sometimes shortcomings of workmanship, rather than inadequate design or specifications, are the cause of unsatisfactory floor performance. Since it is the top surface of the floor that is continually and critically appraised by the user, added attention to the construction of the top surface of the slab and to proper jointing may contribute more to user satisfaction than undue attention to the thickness of the slab itself.

Still, for structural design, a floor thickness must be chosen. Based on extensive data from many years of laboratory and field research and testing from highway pavement engineering, the Portland Cement Association design method for concrete floors on ground is well established. The charts published in this chapter offer a fast means of establishing floor thickness using easily determined parameters. Alternately, computer programs are widely available for floor thickness design.

Plain concrete slabs—ones without distributed steel or structural reinforcement—offer advantages of economy and ease of construction. Acknowledging the obvious similarities and differences between an unreinforced pavement and a plain concrete floor slab, pavement theory was reduced to easily used thickness design charts for floors on ground. The design method was initially presented in *Slab Thickness Design for Industrial Concrete Floors on Grade* (IS195) and is applicable as well to slabs on ground for outdoor storage and material-handling areas. As in pavement design, the factors involved in determining the required floor slab thickness include:

- subgrade and subbase bearing support
- strength of concrete
- location and frequency of imposed loads
- load magnitude, including construction loads

The floor on ground thickness design procedure, as previously published by the Portland Cement Association (IS195), is presented with some modifications in the following material.

Variations in the two concrete properties, modulus of elasticity, E, and Poisson's ratio, μ, have only a slight effect on thickness design. The values used to develop the design charts in this publication are E = 27.6 GPa (4,000,000 psi) and μ = 0.15.

DESIGN OBJECTIVES

Several types of slab distress due to excessive loads can occur—cracking due to excessive flexural stress; excessive deflections; settlement due to excessive soil pressures; and, for very concentrated loads, excessive concrete bearing or shear stresses.

The strategy of floor-slab design is to keep all these responses within safe limits. The most critical of these responses—the controlling design consideration—is different for different sizes of load contact area, as indicated in Fig. 5-1. For example, flexural stress is the controlling

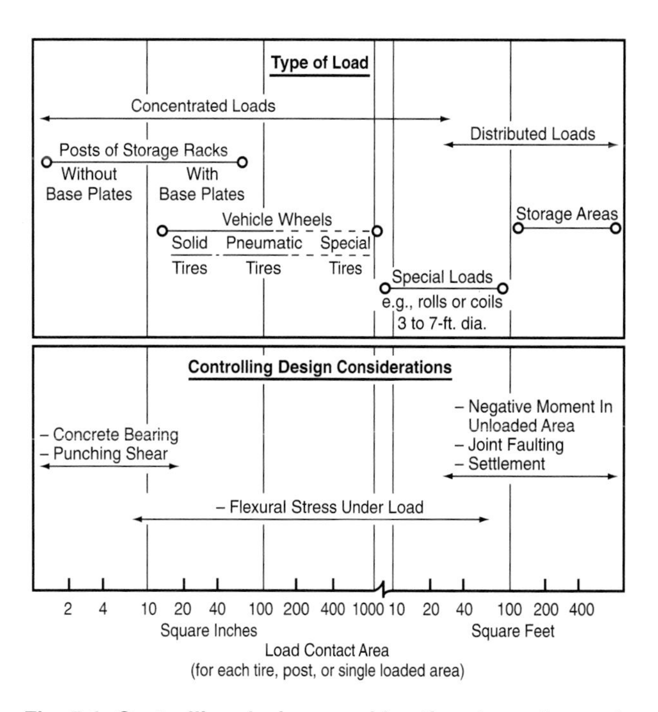

Fig. 5-1. Controlling design consideration depends on size of load contact area.

design consideration for lift trucks that normally have individual wheel contact areas in the range shown. A slab thickness that keeps wheel-load induced flexural stress below a specified safe limit will generally provide an even greater factor of safety against failure from other load responses shown in Fig. 5-1.

For distributed loads covering large areas in storage bays, flexural stress under the load is not as critical as other responses (see Fig. 5-2). Negative moments—tensile stresses in the top of the slab—away from the load may cause a crack in the aisleway, or the load may cause joints to fault as a result of differential settlements. Also, excessive soil pressures due to distributed loads may result in objectionable settlement of some soils.

Load contact area is critical for heavy loads on the leg or post of a storage rack (see Fig. 5-3). If the base plate area is too small, slab distress due to excessive bearing or punching shear is of more concern than the other responses. When the base plate is large enough to prevent a bearing or shear failure, flexural stress becomes the controlling design consideration.

Fig. 5-2. Distributed loads on a floor. (67197)

Fig. 5-3. Post loads on a floor. (69654)

It should be noted that Fig. 5-1 is presented as a guide only. Obviously, boundaries between different controlling design considerations are not exact and will vary somewhat depending on many factors, including slab thickness, concrete strength, and subgrade strength and compressibility. Thus, for values of contact areas between or near the limits shown, the other appropriate responses should be considered in the design.

The load effects and the controlling design considerations are also discussed in the following pages under Vehicle Loads, Post Loads, and Distributed Loads.

FLEXURAL STRESSES AND SAFETY FACTORS

Flexure is a viable design criterion for floors because it directly relates to the overall *structural behavior* of floors under imposed loads (see Fig. 4-7). This strength in bending, flexural strength, is also referred to as the modulus of rupture and abbreviated MOR or MR. Compressive strength relates more directly to the behavior of the *material* (concrete) only. Flexural strength of concrete is proportional to the compressive strength: higher compressive strengths generally yield higher flexural strengths. The modulus of rupture of any concrete can be determined by testing or can be estimated from equations based on the compressive strength (see Chapter 4).

One of the preliminary steps in thickness design of concrete floors is to determine the flexural stress that the concrete can withstand. The allowable working stress is determined by dividing the concrete flexural strength by an appropriate safety factor. The loading on a concrete floor will induce stress in the concrete, and the designer's job is to keep this stress below the allowable stress by choosing the appropriate concrete strength and thickness. The safety factors for vehicle loads have been established based on experience gained in pavement performance and take into account the influence of the number of load repe-

Safety Factors, Shrinkage Stresses, and Impact

Except for long, continuously reinforced slabs, shrinkage stresses are not considered significant. For example, a shrinkage stress of 0.16 MPa (23 psi) is computed for a 200-mm (8-in.) slab jointed at 6 m (20 ft) using the commonly accepted subgrade friction factor of 1.5. Pavement research shows that the actual stress developed will be only a third or half of that computed (Friberg 1954, Kelley 1939).

In some procedures for industrial floor design, the loads are increased by a factor for the effect of wheel impact. However, pavement research shows that slab stresses are less for moving loads than for static loads (HRB 1952, HRB 1962). Therefore, a load impact factor is not used in this procedure.

titions (most important), shrinkage stresses, and impact (see Safety Factors, Shrinkage Stresses, and Impact).

Appropriate safety factors for static loads, either concentrated or distributed, are not well established by experience or research. The designer should give careful consideration to specific design conditions and performance requirements and determine performance characteristics of slabs under similar loading conditions.

Slab stresses for vehicle and post loads were determined by the use of a computer program with appropriate modifications in load contact area (Packard 1967). The flexural stresses indicated in the design charts are those at the interior of a slab, assuming that the load is applied at some distance from any free edge. For loads applied at or near free slab edges, calculated flexural stresses are about 50% to 60% greater than those for interior load positions.* When load transfer occurs across joints (either by dowels or aggregate interlock), flexural stresses at the edge are reduced. This stress reduction depends on stress transfer efficiency.

Because flexural stresses are 50% to 60% greater at slab edges without adequate load transfer, slab thickness should be increased at undoweled butt joints, whether the joints are at the floor periphery or at interior locations. The thickened slab compensates for the absence of load transfer and keeps flexural stresses at these edges within safe limits. See Vehicle Loads (this chapter) for guidelines on thickening slabs.

For butt-type construction joints with dowel load transfer, load-deflection transfer efficiencies are generally 85% or better but the corresponding load-stress transfer efficiencies (calculated) are only about 30% (note: the deflection and stress load transfer efficiency are not the same). Slab thickening aimed at maintaining tolerable flexural stresses is generally not needed for doweled joints. Instead, the working stress is adjusted to accommodate edge stress effects. This is illustrated in the following paragraph.

The slab thickness design charts used in this publication were developed for interior-of-slab load locations. However, the same charts can be used for joints without significant stress transfer efficiency that are subjected to moving loads. Effects of edge stresses are accounted for by adjusting the working stress (*WS*), using an appropriate joint factor (*JF*) based on the higher flexural stresses at loaded edges. For example, for a modulus of rupture of 3.9 MPa (560 psi), using a safety factor (*SF*) of 2.2 for interior load location will provide a working stress of 1.8 MPa (255 psi) (see 5-1 and its accompanying discussion of allowable number of load applications). For edge loads, the *SF* is adjusted by a joint factor of 1.6 (60% higher than for the interior load condition) to account for the higher concrete flexural stresses at the edge. Thus the allowable *WS* is lower: 3.9/(2.2 x 1.6) = 1.1 MPa [560/(2.2 x 1.6) = 160 psi].

*Analysis of concrete flexural stresses using the ILLI SLAB finite element computer program show that edge stresses for free edge loads and interior slab loads range from 50% to 60% greater when compared to interior slab stresses. Calculations were made for lift truck and axle wheel spacings of 910 mm and 1220 mm (36 in. and 48 in.), respectively, and a range of ℓ (radius of relative stiffness) from 510 mm to 1520 mm (20 in. to 60 in.).

When joint load transfer (and thus stress transfer) can be assured—for example, by good aggregate interlock or by dowels—the adjustment to the working stress for frequent loadings is not as great. However, the designer is cautioned that slab contraction caused by cooling and drying may produce control joint cracks that are too wide to permit aggregate interlock load transfer (Tabatabuie, Barenberg, and Smith 1979).

Aggregate interlock load transfer efficiency decreases significantly as the crack widens with time. Crack width depends on the amount of slab contraction and on joint spacings. All cracks are affected, whether they occur beneath contraction joint sawcuts, strip inserts, or at random. The magnitude of contraction depends on concrete placement temperatures relative to operating floor temperatures and on concrete drying shrinkage. *Crack width should not exceed 0.89 mm (0.035 in.) if aggregate interlock is to remain effective* (Colley and Humphrey 1967). For thicker floors, aggregate interlock may be effective even at wider joint openings.

The effects of concrete volume change on contraction joint crack width are illustrated by the example provided in the box, Crack Width Calculation Below a Contraction Joint Sawcut. The coefficient of expansion of concrete is the change in length per unit length with changes in temperature. A typical value for the coefficient of expansion of concrete is 9.9 x 10^{-6}/°C (5.5 x 10^{-6}/°F).

The slab described in the example in the box (Crack Width Calculation) would not have effective load transfer

Crack Width Calculation Below a Contraction Joint Sawcut

Joint spacing = 4.575 m = 4575 mm* (15 ft)

Temperature drop (from placing concrete to floor operation) = 11.1°C (20°F)

Long term drying shrinkage = 200 millionths

ΔL_T Contraction due to temperature = 0.0000099 × Δ T (°C) × L (mm)

ΔL_D Contraction due to drying = (200 ÷ 1,000,000) × L (mm)

Total crack width (in mm) = $\Delta L_T + \Delta L_D$

$$= (0.0000099 \times 11.1 \times 4575) + \left[\left(\frac{200}{1,000,000}\right) \times 4575\right]$$

= 0.502 + 0.915 = 1.417 mm, say 1.42 mm

ΔL_T Contraction due to temperature = 0.0000055 × Δ T (°F) × L (in.)

ΔL_D Contraction due to drying = (200 ÷ 1,000,000) × L (in.)

Total crack width (in.) = $\Delta L_T + \Delta L_D$

$$= (0.0000099 \times 20 \times 15 \times 12) + \left[\left(\frac{200}{1,000,000}\right) \times 15 \times 12\right]$$

= 0.0198 + 0.036 = 0.0558 in., say 0.056 in.

*The convention on metric plans is to use mm for all length measurements to avoid confusion. In this publication, however, length measurements in meters are also used.

at its joints because the crack openings exceed the distance for good aggregate interlock. In order for a slab with this joint spacing and temperature conditioning to have effective aggregate interlock, the drying shrinkage of the concrete would need to be reduced by about one-half. Then, for long-term exposure with zero subgrade restraint, contraction joint crack width would be 0.91 mm (0.036 in.) for the following conditions:

- 4.57 m (15 ft) joint spacing
- 11°C (20°F) temperature drop of the concrete from as-constructed conditions to operating conditions
- drying shrinkage of 100 millionths

VEHICLE LOADS

The design procedure for vehicle loads involves determination of several specific design factors:

- maximum axle load
- number of load repetitions
- tire contact area
- spacing between wheels on heaviest axle
- subgrade-subbase strength
- flexural strength of concrete
- factor of safety
- load transfer at joints

Where long joint spacings are used—certainly anything over 6.0 m (20 ft)—dowels should be considered at contraction joints. Otherwise, loads may not be transmitted across the joint because the opening is too wide. Joint spacings less than 4.5 m (15 ft) are considered rather short and will likely provide good stress transfer. Between these two distances, the designer has some discretion about how good the stress transfer will be. Aggregate interlock effectiveness decreases rapidly as joints open more than 0.89 mm (0.035 in.). When a joint has no provision for load transfer, the slab can be thickened at the joint to improve performance under loading. The typical method is to increase the slab depth by 20% at a taper extending away from the joint a distance 6 to 10 times the pavement thickness (ACPA 1991). Some designers increase the thickness by 25% at a taper not to exceed a 1-in-10 slope. Joint thickening is less common now than it was in the past. For floors with large areas and many joints, concrete placement practicalities make thickened edge joints undesirable.

Floor design requires that traffic be estimated correctly. Traffic information includes:

- load magnitudes
- load frequencies
- wheel configurations of the vehicles riding on the floor

The magnitude of load quantifies the force acting on the floor. Frequency refers to the number of times that a given magnitude of load is applied to the concrete (affected by repeated loading). Failure due to repeated loading is termed *fatigue* and is manifested as cracking. Additionally, the layout of wheels affects how much pressure is placed on the floor. All of these factors together are needed to describe the traffic.

Traffic and load data for past and future facility operating conditions can be gathered from several sources, including plant maintenance and engineering departments, planning and operations departments, and manufacturers' data for lift trucks and other vehicles. Based on this information, an adequate safety factor can be selected to determine an allowable working stress.

For floors, the factor of safety is the ratio of the concrete flexural strength (modulus of rupture) to the working flexural stress. It can be thought of as a ratio of the total capacity available before failure would occur to the amount of strength being used. The inverse of safety factor (flexural stress divided by flexural strength) is called the stress ratio. In fatigue studies, stress ratio values are related to allowable load repetitions.

As long as the stress ratio is kept at or below 0.45, concrete can endure an unlimited number of load repetitions without fatigue cracking. (A stress ratio of 0.45 is equivalent to a safety factor of 2.2.) For stress ratios greater than 0.45 (factor of safety less than 2.2), Table 5-1 lists the maximum number of load repetitions that can be sustained without causing fatigue cracking.

Table 5-1. Stress Ratio Versus Allowable Load Repetitions (PCA Fatigue Curve)*

Stress ratio	Allowable load repetitions	Stress ratio	Allowable load repetitions
< 0.45	unlimited	0.73	832
0.45	62,790,761	0.74	630
0.46	14,335,236	0.75	477
0.47	5,202,474	0.76	361
0.48	2,402,754	0.77	274
0.49	1,286,914	0.78	207
0.50	762,043	0.79	157
0.51	485,184	0.80	119
0.52	326,334	0.81	90
0.53	229,127	0.82	68
0.54	166,533	0.83	52
0.55	124,523	0.84	39
0.56	94,065	0.85	30
0.57	71,229	0.86	22
0.58	53,937	0.87	17
0.59	40,842	0.88	13
0.60	30,927	0.89	10
0.61	23,419	0.90	7
0.62	17,733	0.91	6
0.63	13,428	0.92	4
0.64	10,168	0.93	3
0.65	7,700	0.94	2
0.66	5,830	0.95	2
0.67	4,415	0.96	1
0.68	3,343	0.97	1
0.69	2,532	0.98	1
0.70	1,917	0.99	1
0.71	1,452	1.00	0
0.72	1,099	>1.00	0

**Thickness Design for Concrete Highway and Street Pavements*, EB109.01P, Portland Cement Association, Skokie, IL, 1984.

The safety factor or corresponding stress ratio depends on the frequency of the heaviest lift-truck loading. Safety factors of 2.2 can be used for entire facilities; in the case of large facilities, it might be economical to choose variable safety factors. Wherever a large number of load repetitions is expected, a high safety factor can be used (usually 2.0 or higher). In other areas where less traffic is expected, a lower safety factor can be chosen. For truck aisles it might be 1.7 to 2.0. For non-critical areas, like uniformly loaded storage areas, it might be 1.4 to 1.7 (Packard and Spears). This design flexibility can save money on concrete materials by allowing for thinner floor sections or lower strength concrete mixtures. However, reduced thickness areas may limit future alternate floor-use strategies.

Because of the large variety of sizes, axle loads, and wheel spacings of industrial trucks, it is not practical to provide separate design charts for each vehicle. Consequently, two design charts, Figs. 5-4 and 5-5, have been prepared and can be used for the axle loads and axle-wheel configurations of most industrial trucks affecting floor design. (Important note: Presenting these thickness design charts in dual units would be cumbersome and confusing. Instead, units should be converted from metric to in.-lb before proceeding with thickness design by this method.)

Fig. 5-4 is used for industrial trucks with axles equipped with single wheels. The chart is entered with an allowable working stress per 1,000 lb of axle load. This allowable stress is computed by dividing the concrete flexural strength by the safety factor and, if needed, by the joint factor and then dividing this result by the axle load in kips (1 kip = 1,000 lb). The safety factor is obtained from considerations of the stress ratio and load repetitions, as provided in Table 5-1.

For axles equipped with dual wheels, Figs. 5-4 and 5-5 are used together to determine floor slab thickness. First, Fig. 5-5 is used to convert the dual-wheel axle load to an equivalent single-wheel axle load (the axle load is multi-

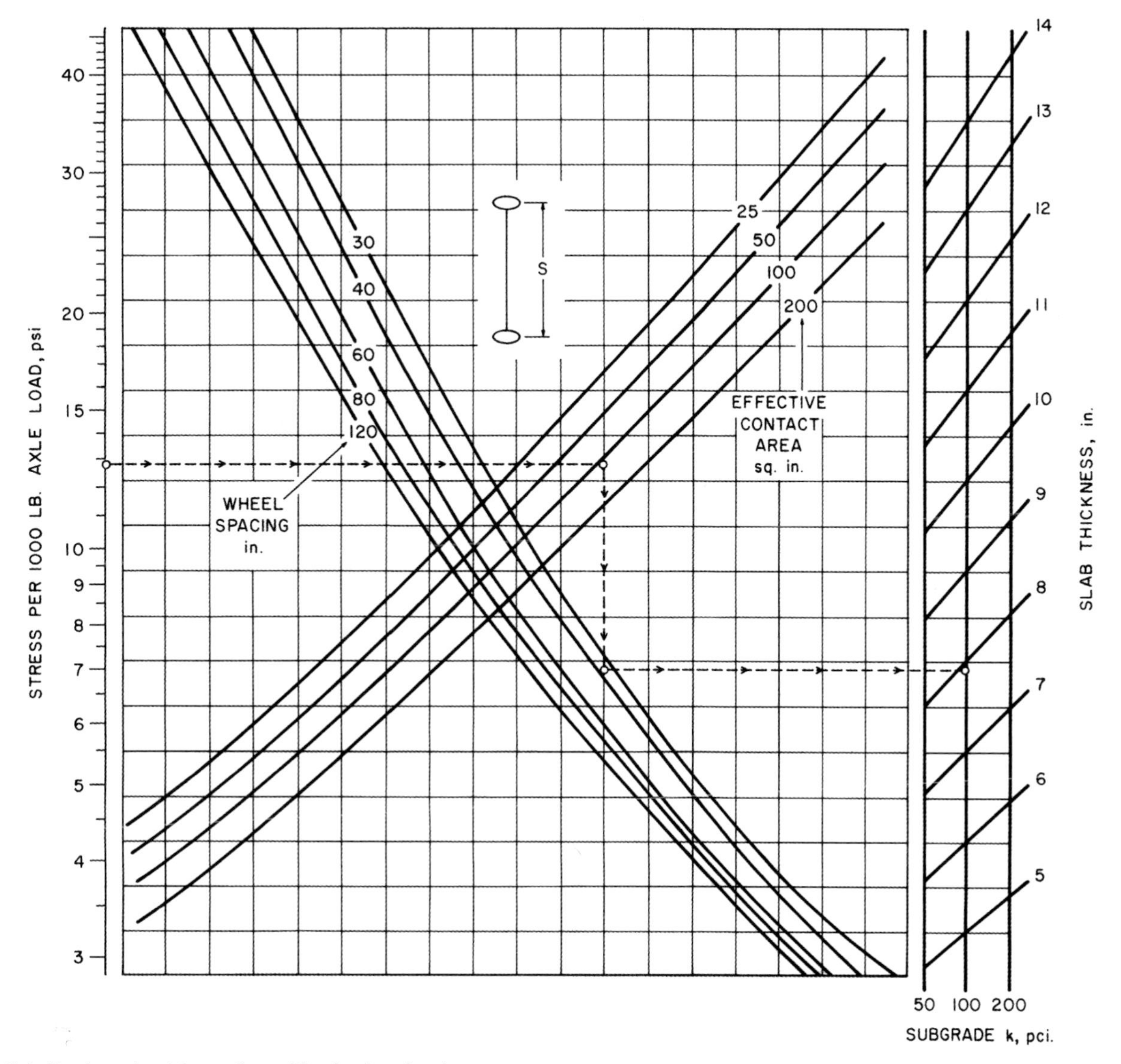

Fig. 5-4. Design chart for axles with single wheels.

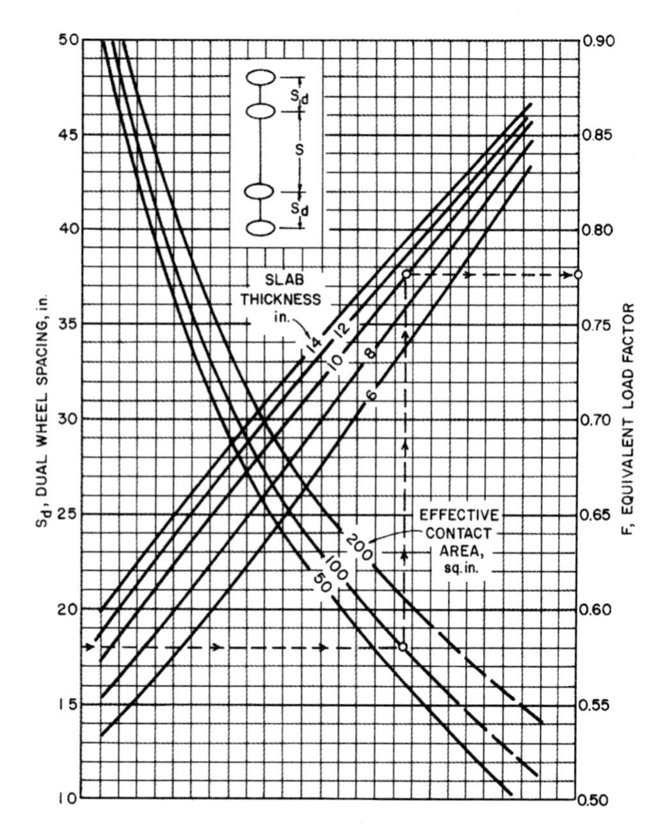

Fig. 5-5. Design chart for axles with dual wheels.

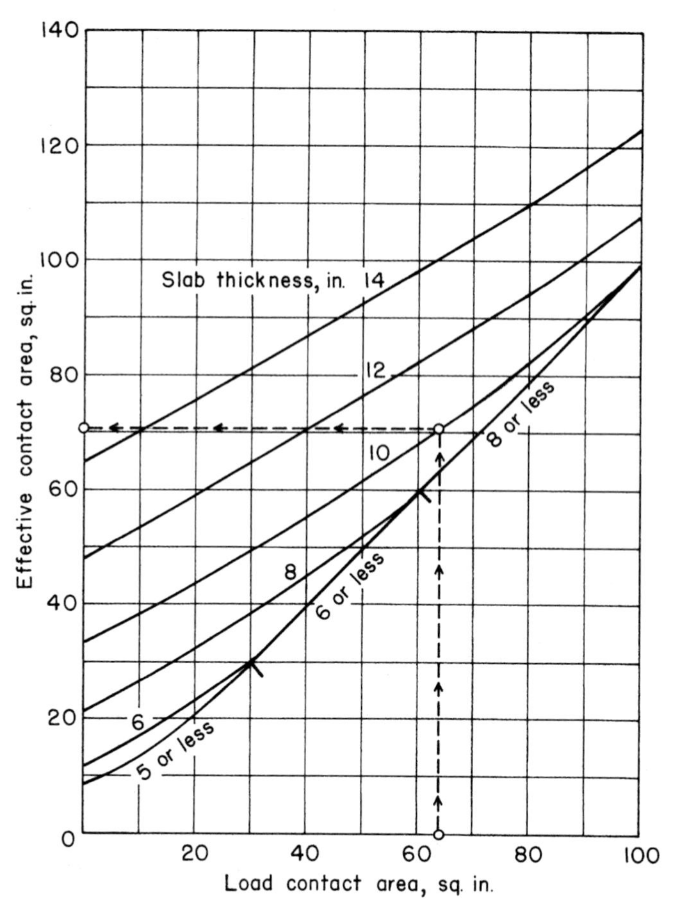

Fig. 5-6. Effective load contact area depends on slab thickness.

plied by the factor *F*). Then, with the equivalent load, Fig. 5-4 is used to determine the flexural stress in the slab.

The load contact area refers to the gross contact area of one tire against the slab, regardless of the tire tread design. If tire data are not available, the contact area may be estimated for pneumatic tires by dividing wheel load by inflation pressure, and roughly approximated for solid or cushion tires by multiplying tire width (in.) by three or four. If the tire size is known, the tire data may be obtained from manufacturers' tables (Tire and Rim Association 1974 and Goodyear 2001).

When the tire contact area has been determined, Fig. 5-6 is used to find the effective contact area for use in the design charts. This correction is made because slab stresses for small load contact areas are overestimated when computed by conventional theory. The basis for this adjustment is given in Westergaard 1925. (This same adjustment is used for post loads discussed in a later section.) In using Fig. 5-6 it is necessary to assume a slab thickness; this is a trial-and-error process to be checked against the final required design thickness. The degree of correction increases as contact area becomes smaller and slab thickness becomes greater.

DESIGN EXAMPLE—VEHICLE LOADS, SINGLE WHEEL

The following example problems illustrate the use of Figs. 5-4 and 5-5 for slab-thickness design for vehicle loads: NOTE: CONVERT ALL METRIC VALUES TO IN.-LB VALUES.

Data for Lift Truck A

Axle load	111 kN (25 kips)
Wheel spacing	940 mm (37 in.)
No. of wheels on axle	2
Tire inflation pressure	0.76 MPa (110 psi)

$$\text{Tire contact area} = \frac{\text{wheel load}}{\text{inflation pressure}}$$

$$= \left(\frac{\frac{25{,}000}{2}}{110}\right) = 114 \text{ sq. in.}$$

Subgrade and Concrete Data

Subgrade modulus, k	27 MPa/m (100 pci)
Concrete flexural strength, MR	4.4 MPa (640 psi) at 28 days

Design Steps (Convert all metric values to in.-lb values before continuing)

1. Safety factor:

Select the safety factor using the data from Table 5-1 for the expected number of forklift loads in channelized aisle traffic. Select 2.2 for unlimited stress repetitions. If anticipated load repetitions are expected to number about 750,000, use 2.0. (*SF* is the inverse of stress ratio.)

2. Joint Factor:

For interior load-based design the joint factor is 1.0. For frequent load crossings of joints without load (or stress) transfer, the *JF* is 1.6. This value is used because floor slab joint spacing at this example facility will be greater than 4.5 m (15 ft). Thus aggregate interlock load transfer will diminish with load repetitions. For many facilities, joint spacings are at 1/2 or 1/3 of column spacings. Joint spacings less than 4.5 m (15 ft) are not often used, even though these closer spacings can permit decreased floor thickness.

3. Concrete working stress:

$$\mathrm{WS} = \left(\frac{\mathrm{MR}}{\mathrm{SF} \times \mathrm{JF}}\right) = \left(\frac{640}{2.2 \times 1.6}\right) = 182 \text{ psi}$$

4. Slab stress per 1,000 lb of axle load:

$$= \left(\frac{\mathrm{WS}}{\text{axleload, kips}}\right) = \left(\frac{182}{25}\right) = 7.3 \text{ psi}$$

5. Enter the left-hand axis of Fig. 5-4 with a stress of 7.3 psi and move right to contact area of 114 in.^{2} From that point, proceed upwards to a wheel spacing of 37 in. From there, move horizontally to the right to read slab thickness of 11.2 in. on the line for subgrade k = 100 pci and use a 11-1/4 in.-thick slab.

If lower use areas are identifiable (large facilities), choose safety factors from Table 5-1 for the estimated number of load repetitions. This may result in reduced floor thickness for those areas.

DESIGN EXAMPLE—VEHICLE LOADS, DUAL WHEELS

Data for Lift Truck B

Axle load	222.4 kN (50 kips)
Dual wheel spacing	460 mm (18 in.)
Wheel assembly spacing	1015 mm (40 in.)
No. of wheels on axle	4
Tire inflation pressure	0.86 MPa (125 psi)

$$\text{Tire contact area} = \frac{\text{wheel load}}{\text{inflation pressure}}$$

$$= \left(\frac{\frac{50{,}000}{4}}{125}\right) = 100 \text{ sq. in.}$$

This contact area is large enough that correction by the use of Fig. 5-6 is not required.

Subgrade and Concrete Data

Subgrade modulus, k	27 MPa/m (100 pci)
Concrete flexural strength, MR	4.4 MPa (640 psi) at 28 days

Design Steps

1. Safety factor:

Lift truck B will carry its maximum load inside the warehouse infrequently, only once or twice a week. For a 40-year design life, there will be about 4000 load repetitions. Thus Table 5-1 provides a stress ratio of 0.67 and a safety factor of 1.5.

2. Joint factor:

With anticipated joint spacing at 1/3 of column spacing (3.5 m) (12 ft), the joint factor can be taken as 1.3 if aggregate interlock (and stress transfer) is good.

3. Concrete working stress:

$$\mathrm{WS} = \left(\frac{\mathrm{MR}}{\mathrm{SF} \times \mathrm{JF}}\right) = \left(\frac{640}{1.5 \times 1.3}\right) = 328 \text{ psi}$$

4. Enter Fig. 5-5 with a dual wheel spacing of 18 in.; move right to a contact area of 100 in.^{2}; then up to a trial slab thickness of 10 in.; then right to an equivalent load factor, F, of 0.775. The equivalent single-wheel axle load is the factor F times the dual-wheel axle load = 0.775 x 50 = 38.8 kips. In using these figures, it is necessary to assume a trial slab thickness to do the graphical solution. The result (design thickness) will have to be compared against the assumed thickness. This trial-and-error process—steps 3 through 5—may have to be repeated until the assumed thickness and the design thickness agree.

5. Slab stress per 1,000 lb of axle load

$$= \left(\frac{\mathrm{WS}}{\text{axleload, kips}}\right) = \left(\frac{328}{38.8}\right) = 8.5 \text{ psi}$$

6. Enter Fig. 5-4 with stress of 8.5 psi; move right to contact area of 100 in.^{2}; move up to a wheel spacing of 40 in.; move right to a slab thickness of 10.3 in. on the line for subgrade *k* of 100 pci. Use a 10.5-in. slab thickness. The 10.5-in. slab thickness is about the same thickness as was assumed converting dual wheels to equivalent single wheel loads. Thus steps 4 through 6 need not be repeated.

7. Convert thickness back to metric equivalent (if needed):

10.5 in. = 265 mm (rounded to nearest 5 mm)

In preliminary design stages, or when detailed design data are not available, Fig. 5-7 may be used as a guide to indicate slab thickness based on the rated capacity of the heaviest lift trucks that will use the floor. This preliminary design will be based on interior slab load location. To adjust thickness for joints with inadequate stress transfer, increase preliminary thickness by 20%. The figure was prepared for typical lift trucks from manufacturers' data, composites of which are shown in Table 5-2 (see Table 5-2 inside back cover). The figure would not apply for vehicles with load and wheel-spacing data that differ substantially from the tabulated data.

The conservative assumptions in Fig. 5-7 regarding the subgrade strength and working stress in the concrete should be noted. The combination of these assumptions results in a greater-than-usual degree of conservatism, which seems necessary when detailed design data are not available. Fig. 5-7 is intended as a rough guide only; more reliable and usually more economical designs may be obtained using more complete design data and Figs. 5-4, 5-5, and 5-6.

POST LOADS

In many industrial buildings and warehouses, racks are used for storing products or materials. If the rack loads are heavy, posts supporting the rack induce significant stresses in the floor slab. Flexural stresses from these concentrated loads can be greater than stresses caused by wheel loads of vehicles operating in the building, and thus may control the thickness design of the floor slab. High rack storage and the high-lift forklift trucks that serve them necessitate close tolerances in floor smoothness. There is a great need to eliminate or at least minimize slab cracking from loading to provide a good riding surface for vehicle operation.

Bearing Plates and Flexural Stresses

For inadequate-size base plates, concrete bearing and shear stresses may be excessive even though flexural stresses are not. The size of the base plate should be large enough so that concrete bearing stress under maximum service load does not exceed 4.2 times the 28-day modulus of rupture, or half of this for loads applied at slab edges or corners. With an adequate base plate size to control bearing stresses and an adequate slab thickness to control flexural stresses, shear stresses are not excessive for the ranges of design variables indicated in this section.

In this case, the allowable shear is taken as 0.27 times the modulus of rupture and the critical section for shear is assumed to act at a distance of half the slab depth from the periphery of the loaded area. For loads at an edge or corner of the slab, any section along a joint is excluded. These criteria are a suggested interpretation of how building code requirements (ACI 318) may be applied to the situation of post loads on floor slabs.

For post loads, the design objective is to keep flexural stresses in the slab within safe limits. Within the range of design variables presented in this section, flexure controls the slab-thickness design. When flexural requirements are satisfied with an adequate slab thickness, soil pressures are not excessive; and when the appropriate size of base plate is used, concrete bearing and shear stresses are not excessive (see box, Bearing Plates and Flexural Stresses).

Because flexure controls, the design factors are similar to those used for vehicle loads. An even higher safety factor may be appropriate, however, if the rack posts are also used to transmit roof loads to the floor slab in place of an independent column footing. The specific design factors are:

- maximum post load
- load contact area
- spacing between posts
- subgrade-subbase strength
- flexural strength of concrete
- safety factor

Figs. 5-8a, 5-8b, and 5-8c are used to determine the slab thickness requirements for k-values of 50, 100, and 200 pci. The charts were developed to estimate interi-

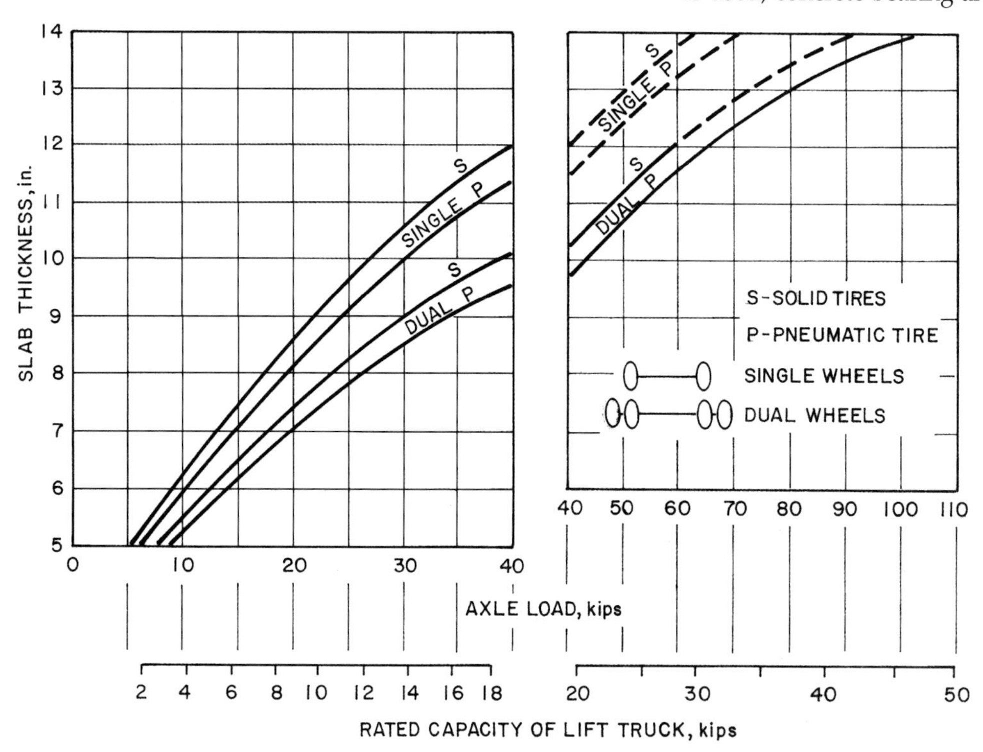

Fig. 5-7. Estimated slab thickness for lift trucks (based on average truck data shown in Table 5-2 and conservative design assumptions: k = 50 pci; concrete working stress of 250 psi).

or slab stresses for the two equivalent post configurations and load conditions shown schematically in Fig. 5-9, representing continuous racks. Fig. 5-10 shows a similar schematic for edge-of-slab loading. (For a structurally reinforced slab, bending moments computed from the flexural stress determined from Figs. 5-8a, 5-8b, and 5-8c may be used to compute the required tensile reinforcement.) Edge loading places higher stresses on floors and, unless accounted for, could lead to more cracking. One way to address higher stress is to modify the working stress values with a joint factor before entering Figs. 5-8.

Another option for addressing higher edge stresses is to use supplemental charts that have been developed for post loads along slab edges (Okamoto and Nussbaum 1984). Figs. 5-11a, 5-11b, 5-11c, and 5-11d take into account the higher edge stress (the joint factor is incorporated into the charts) and are used in much the same way as Figs. 5-8. In Figs. 5-11, the stress on the post load is calculated, x- and y-spacings are determined, the correct chart is entered (for the y-spacing), and the subgrade modulus is used to help read the correct slab thickness off the right-hand side of the chart. If the base plate differs from a contact area of 10 in.^{2}, the allowable stress should be modified to a design stress using chart 5-12a or 5-12b for Figs. 5-11. One is for tensile stress in the bottom edge and the other is for tensile stress in the top edge. Figures 5-11 and 5-12 are currently only available in U. S. customary units, so conversions from metric should be made before entering the charts and then again to convert the final results back to metric. Similarly, Fig. 5-6 provides the correction for the load contact area to effective contact area for design charts in Figs. 5-8. Figs. 5-8 and 5-11 should yield the same result (the same slab thickness) for the same in-put (post configuration, loads, sub-grade modulus, and contact area).

In Figs. 5-8 and 5-11, the post spacing, y, is in the longitudinal direction of a continuous rack and x is the transverse spacing. The charts provide spacings from 40 in. to 100 in. in Fig. 5-8 and from 20 in. to 100 in. in Fig. 5-11. Intermediate spacings can be interpolated in these figures. If two posts come close enough together so that their base plates would touch or overlap (e.g. back-to-back racks), the posts can be assumed to act as one load equivalent to the sum of their combined loads. The values in these charts represent common rack spacings. Computer programs allow more flexibility in choosing the distances between posts.

The k-values represent generic soil conditions for low, medium, and somewhat high soil support. In these design charts, the k-values are fixed, though it is possible to interpolate between them with acceptable results.

These thickness design charts apply to a wide range of slabs. Because they cannot cover all situations, it may be necessary to use other means to determine slab thickness. Computer programs allow for input parameters of all aspects of the slab design, including soil support, special

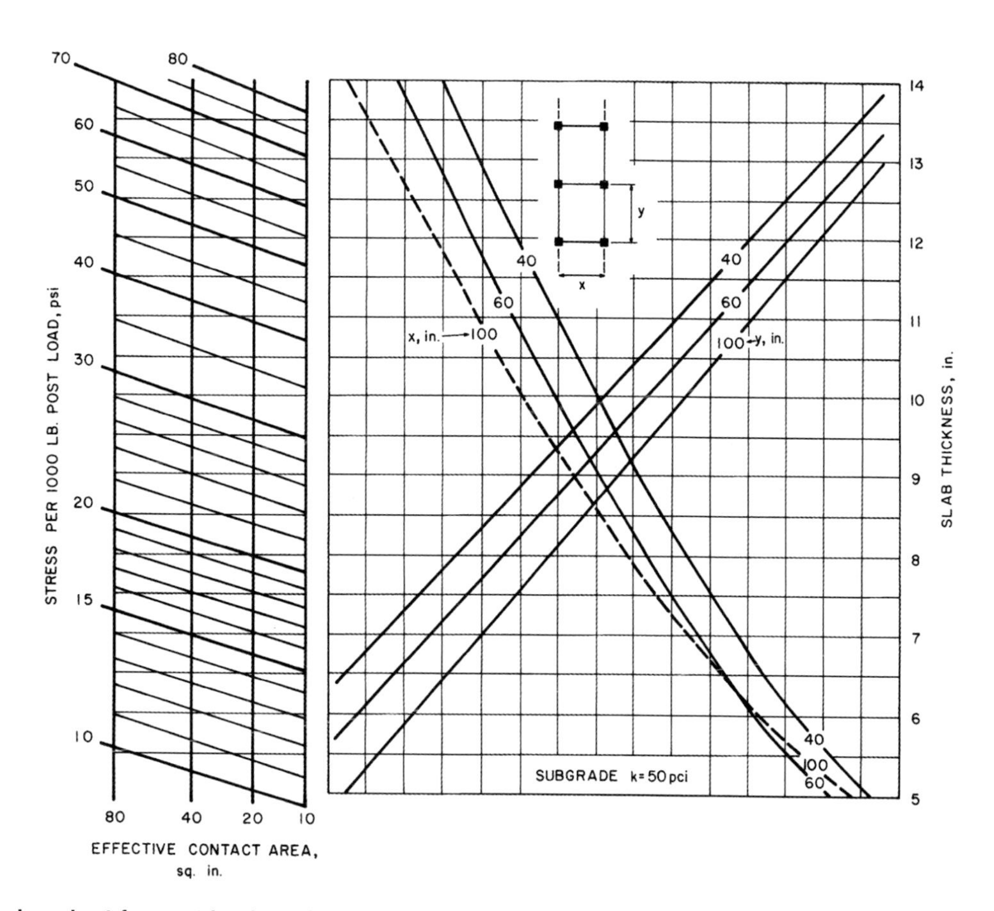

Fig. 5-8a. Design chart for post loads, subgrade k = 50 pci.

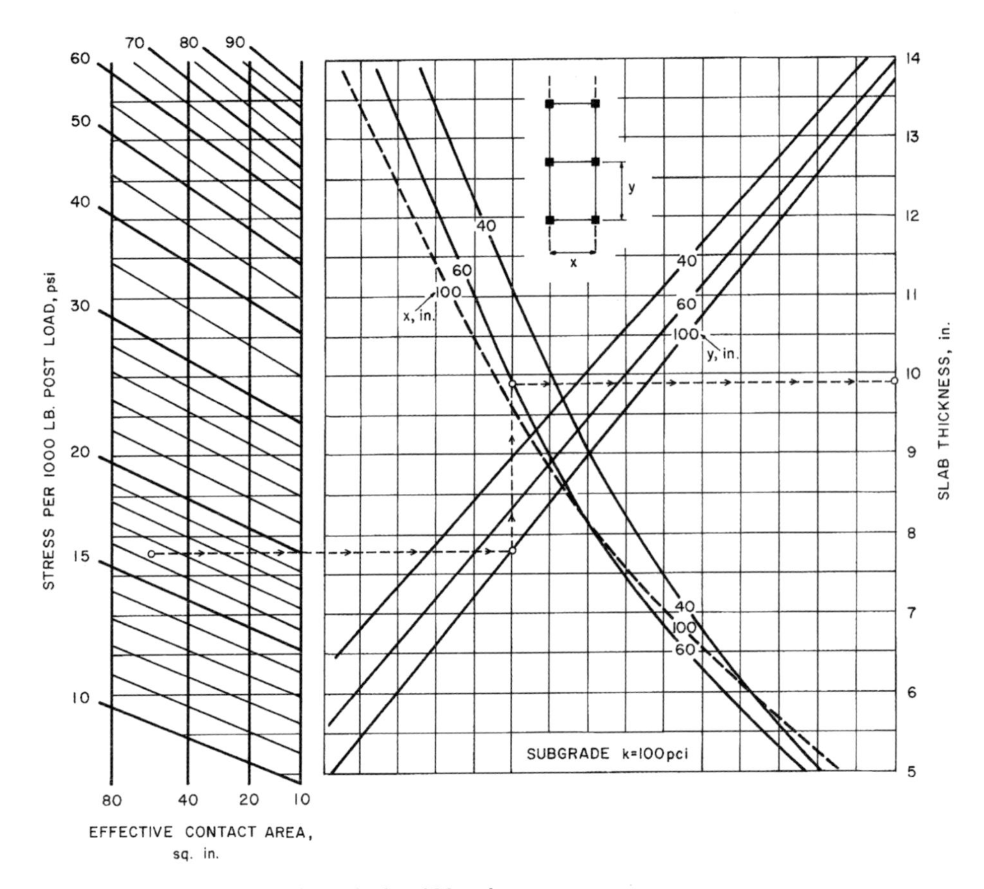

Fig. 5-8b. Design chart for post loads, subgrade k = 100 pci.

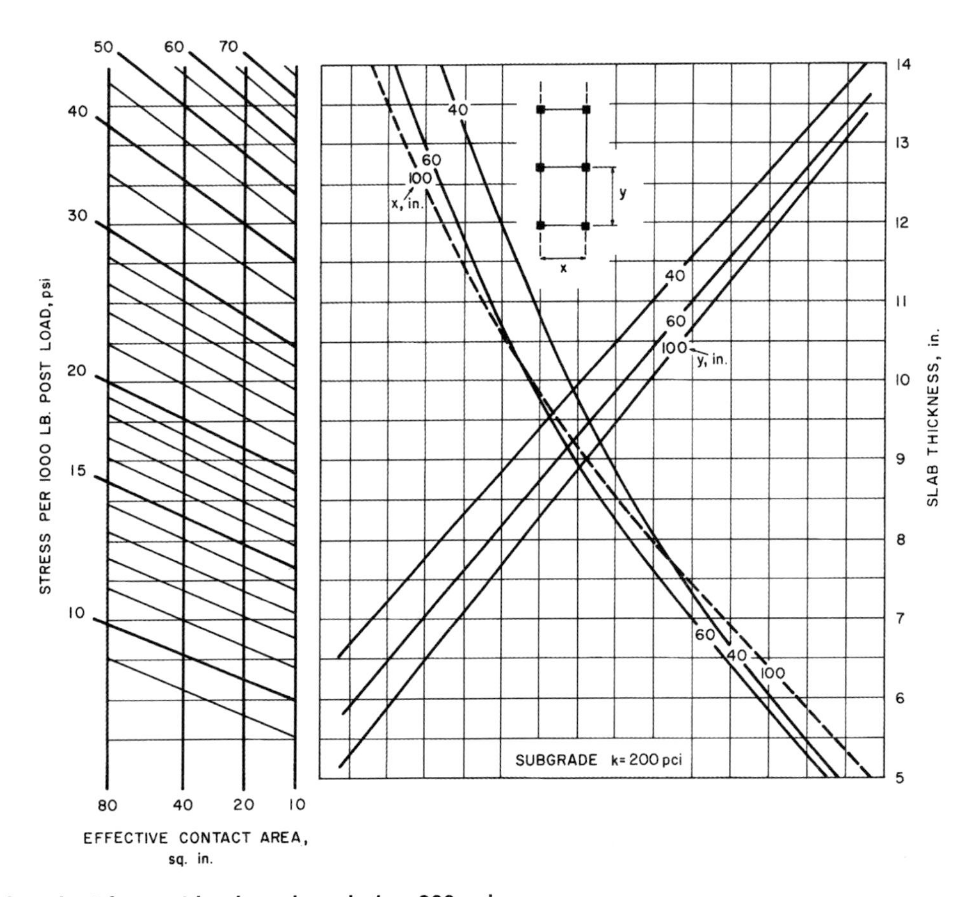

Fig. 5-8c. Design chart for post loads, subgrade k = 200 pci.

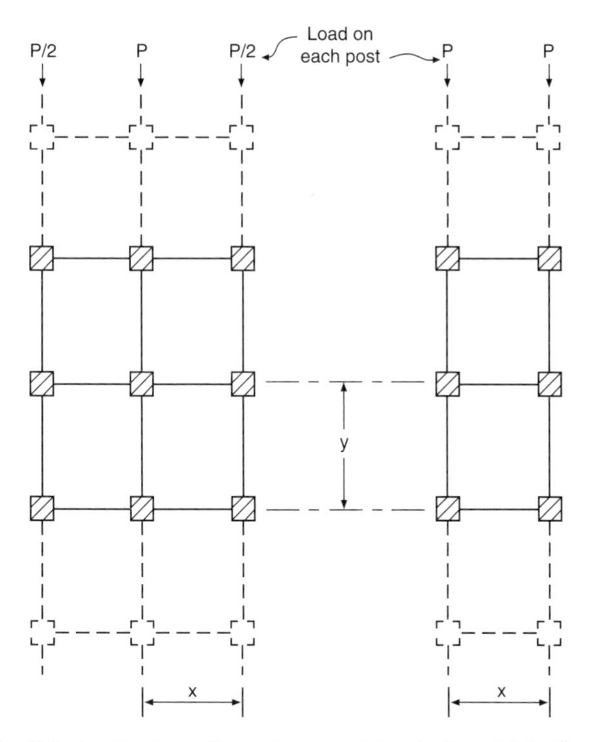

Fig. 5-9a,b. Post configurations and loads for which Figs. 5-8a, 5-8b, and 5-8c apply (interior loading). (69655)

load configurations, and concrete strength. For special post load configurations that deviate substantially from those indicated in Fig. 5-9, slab stresses may be determined by computer program (ACPA 1992) or by influence charts (Pickett and Ray 1951). The computer program may be used with appropriate modifications in the shape of the contact area. For the range of contact areas involved, a circular or elliptical area may be used without significant error to approximate a square or rectangular area.

It should be noted that the design procedure is based on load stresses only; it is not necessary to consider shrinkage stresses (see box, Safety Factors, Shrinkage Stresses, and Impact).

Safety Factors for Post Loads

The specific safety factors to be selected for concentrated static loads are not given here but are left to the judgment of the design engineer. There are two reasons for this:

1. The range of possible safety factors may be quite wide; the factor may be relatively low—1.5 or less—under a non-critical loading condition, or quite high—approximately 5—in a situation where consequences of slab failure are quite serious.
2. Performance experience and experimental data for concentrated static loads are not available.

Some of the points to be considered in the selection of the safety factors are discussed below.

Static loads on posts have effects different from loads on vehicles in that (1) moving wheel loads produce lower slab stresses than static loads of the same magnitude, and (2) creep effects reduce stress under static load. Information on how these effects may be quantified in design problems is not available.

There are reasons to use higher safety factors for loads on high racks than those used for low racks, vehicle loads, or distributed loads. The rack posts are sometimes designed to partially support the roof structure, and effects of differences in deflection between rack posts are magnified with high racks.

Since there is a lack of published data on performance experience with rack loads on slabs on grade, safety factors cannot be suggested with as much confidence as for vehicle loads. This makes it important to carefully consider the characteristics of this type of loading and the desired performance requirements.

A safety factor of 4.8 can be used based on building code requirements if the post is considered a critical structural element—a column—in the building and the slab is considered an unreinforced spread footing (ACI 318).

This value of 4.8 is considered the upper limit of the safety factor range because the post load situation is usually not as critical as that for columns on footings. Columns are spaced farther apart and each supports a greater proportion of the total structural load. The fundamental difference between the two types of loading lies in differences in pressures on the underlying soil. Soil pressures under a footing may be near the limit of allowable soil bearing; if a

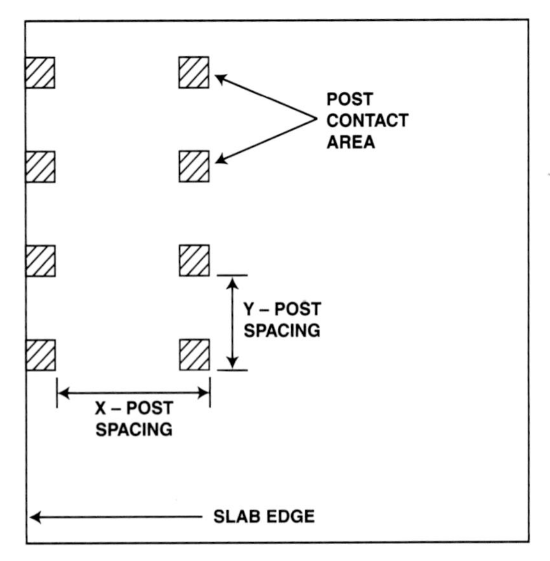

Fig. 5-10. Post configurations for post loading along the slab edge.

failure should occur in the footing, the allowable soil pressure would be exceeded and there would be a possibility of intolerable soil penetration, settlement, or complete collapse. On the other hand, soil pressures under a slab of adequate thickness supporting a post load are much lower than those for a footing. This is because the slab distributes the load over a large area of subgrade. Even if a joint or crack (or intersection of either) should occur at a post, deflections and soil pressures will be increased by a magnitude of 2 or 3 but are still not excessive (Childs 1964). Soil pressures can be computed from experimental data from deflections due to loads at slab corners, edges, and interiors.

Joint Factors for Post Loads

For warehouses and storage areas of manufacturing facilities, storage racks are frequently placed back-to-back along, and parallel to, column lines.

The column line alignments often coincide with construction or contraction joints. If joint spacings are short or if doweled joints are used, effective load transfer occurs. Joint spacings are considered short if they are less than about 4.5 m (15 ft). With good load transfer, edge slab stresses are reduced 20% to 25% along the jointed edges (Okamoto and Nussbaum 1984). This effect can be accounted for by increasing the working stress by 20%. If longer joint spacings or no dowels are used, aggregate interlock (and stress transfer) across sawcut joints is generally ineffective, and a joint factor of 1.6 should be used for rack-post-loading slab thickness determination. (Note: keyed construction joints should not be used.)

Calculate the working stress, *WS*, by dividing the concrete modulus of rupture by the product of the safety factor and joint factor. Thus, if needed, Figs. 5-8a, 5-8b, and 5-8c for interior slab loading conditions can be used for post-load edge loading locations when the joint factor is used for calculating working stress.

After the designer has selected an appropriate safety factor and joint factor based on the criticality of the loading conditions, Fig. 5-8a, 5-8b, or 5-8c is used to establish a slab design thickness based on flexure. Shear stress and concrete bearing stress should also be computed to determine if these values are within safe limits. The following example problem illustrates the procedure for determining slab stresses due to post loads.

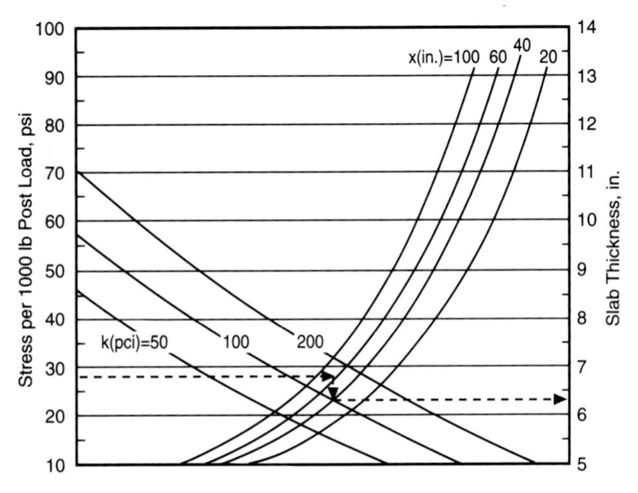

Fig. 5-11a. Design chart for slab edge post loads, post-spacing: y = 20 in., stress at slab bottom surface.

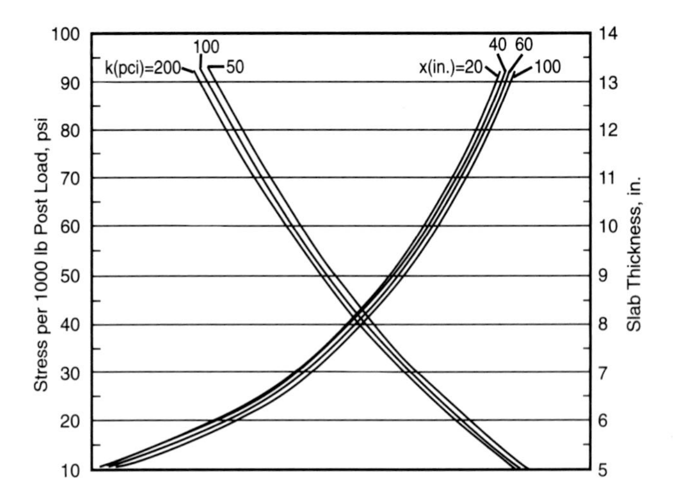

Fig. 5-11b. Design chart for slab edge post loads, post-spacing: y = 40 in., stress at slab bottom surface.

DESIGN EXAMPLE—POST LOADS

Data for Post Configuration and Load

CONVERT ALL METRIC VALUES TO IN.-LB VALUES.

Post spacing	longitudinal (y): 2500 mm (98 in.)
	transverse (x): 1700 mm (66 in.)
Post loads	57.8 kN (13 kips), each post
Load contact area	41,210 mm² (64 in.²), a 203-mm (8-in.) square plate

Subgrade and Concrete Data

Subgrade modulus, k	27 MPa/m (100 pci)
Concrete flexural strength, MR	4.4 MPa (640 psi) at 28 days

Floor Joint Spacing

Column spacing	15200 mm (50 ft)
Construction and/or contraction joint spacing	5100 mm (16.7 ft)

Design Steps

1. Safety factor:
   Select SF of 2, as posts carry no building components and rack height is less than 10.7 m (35 ft).

2. Joint factor:
   Select JF of 1.6, as joint spacing is in excess of 4.5 m (15 ft) and aggregate interlock joint stress transfer consequently is negligible.

3. Concrete working stress:

$$WS = \left(\frac{MR}{SF \times JF}\right) = \left(\frac{640}{2 \times 1.6}\right) = 200 \text{ psi}$$

4. Slab stress per 1,000 lb of post load:

$$= \left(\frac{WS}{\text{post load, kips}}\right) = \left(\frac{200}{13}\right) = 15.4 \text{ psi}$$

5. For a subgrade k-value of 100 pci, use Fig. 5-8b. In the grid at the left of the figure, locate the point corresponding to 15.4 psi stress and 64 in.² contact area. Then move right to y-post spacing of 98 in. Move up to x-post spacing of 66 in. Then move right to a slab thickness of 10.4 in. (use a 10.5 in. thick slab).

6. Use Fig. 5-6 to determine if the effective contact area is significantly larger than the actual contact area. For a 10.5-in. slab and contact area of 64 in.², the effective contact area is 72 in.²; this correction does not significantly change the required slab thickness.

7. The following check of concrete bearing and shear stresses indicates that they are within allowable limits (see Bearing Plates and Flexural Stresses, page 37).

Allowable bearing stress

At 4.2 times the 28-day modulus of rupture for interior load
$= 4.2MR = 2{,}690$ psi

At 1/2 of 4.2 times the 28-day modulus of rupture for edge or corner load
$= 2.1MR = 1{,}345$ psi

Computed bearing stress =

$$\frac{\text{post load}}{\text{load area}} = \frac{13{,}000}{64} = 203 \text{ psi}$$

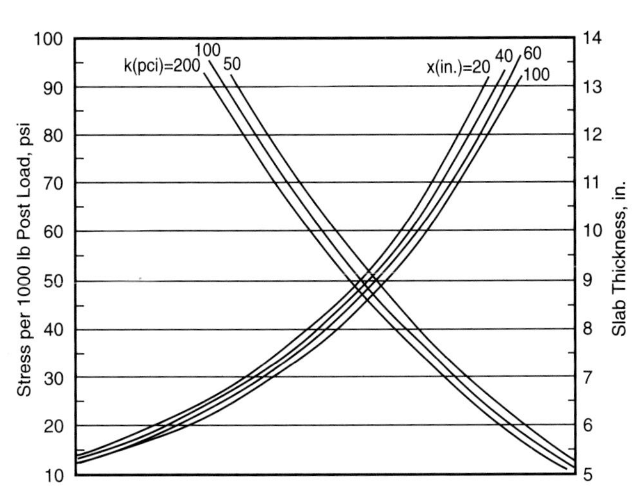

Fig. 5-11c. Design chart for slab edge post loads, post-spacing: y = 60 in., stress at slab bottom surface.

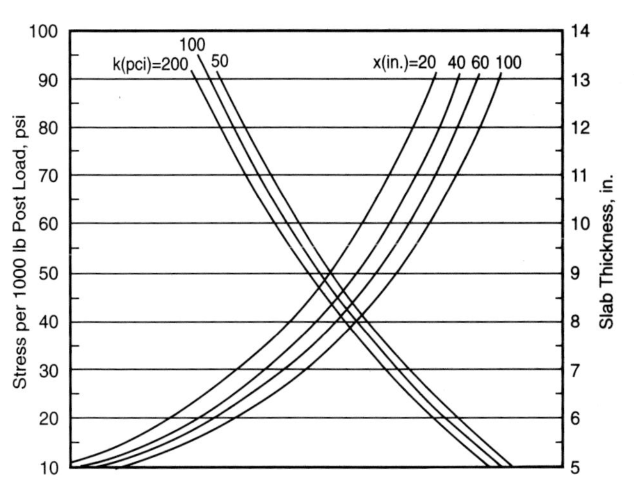

Fig. 5-11d. Design chart for slab edge post loads, post-spacing: y = 100 in., stress at slab top surface.

Bearing stress is significantly less than allowable bearing stress.

Allowable shear stress

$= 0.27 \times MR = 0.27 \times 640 = 173$ psi

$$\text{Computed shear stress} = \frac{\text{post load}}{\text{shear area}}$$

Computed shear stress for interior load

Load periphery is four times the square root of

$$= \frac{\text{post load}}{\text{slab thickness} \times [\text{load periphery} + (4 \times \text{slab thickness})]}$$

Load periphery is four times the square root of bearing area — here :

$= 4\sqrt{64} = 32$ in.

So the computed shear stress for the bearing area is:

$$= \frac{13{,}000}{10.5 \times [32 + 42]} = 17 \text{ psi}$$

Computed shear stress for edge load:

$$= \frac{\text{post load}}{\text{slab depth} \times [(0.75 \times \text{load periphery}) + (2 \times \text{slab depth})]}$$

$$= \frac{13{,}000}{10.5 \times [(0.75 \times 32) + (2 \times 10.5)]} = 28 \text{ psi}$$

Computed shear stress for corner load:

$$= \frac{\text{post load}}{\text{slab depth} \times [(0.5 \times \text{load periphery}) + (\text{slab depth})]}$$

$$= \frac{13{,}000}{10.5 \times [(0.5 \times 32) + (10.5)]} = 47 \text{ psi}$$

Calculated corner, edge, and interior shear stresses are significantly less than allowable concrete shear stresses.

Convert thickness of slab back to metric (if needed): 10.5 in. = 265 mm

Alternately, the floor thickness design for post loads located parallel to and along a slab edge or joint with no load transfer may be accomplished through the use of Figs. 5-10 through 5-12. Post load positions are shown in Fig. 5-10. Figs. 5-11a through 5-11d were based on analysis of the effect of post loading along slab edges using the JSLAB finite element analysis program. Figs. 5-11a, b, and c are for stresses at slab bottom. Fig. 5-11d is for stresses at slab top surface (negative bending stress). The following inputs (in.-lb) were used for the analysis:

E = 4 million psi
Poisson's ratio = 0.15
Slab dimensions = 17 ft by 17 ft
Post contact area = 10 in.^{2}
Modulus of subgrade reaction, *k*, of 50, 100 and 200 pci
Four spacings of post loads in each direction

Joint factors are not included as the design is based on free edge loading. Taking an example, for post load spacings of y = 20 in., x = 60 in., and k = 100 pci, use Fig. 5-11a. (For other y-dimensions, use Figs. 5-11b, c, or d.) With roughly 28-psi stress per 1000 lb post load for the example shown, move horizontally to the x = 60 in. line, drop a vertical line to the k = 100 pci curve, then move horizontally to the right axis to determine a slab thickness of 6.4 in.

For a contact area other than 10 in.^{2} a correction factor is used to account for the beneficial effect of a larger bearing plate under the post. Correction factors are provided in Fig. 5-12a for Figs. 5-11a, b, and c (for slab bottom surface

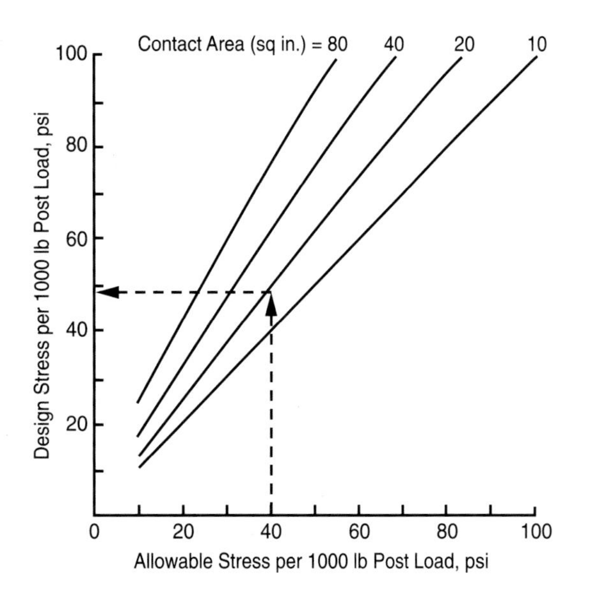

Fig. 5-12a. Design stress depends on allowable stress and contact area for the case of tensile stress at the slab bottom for slab edge post loads.

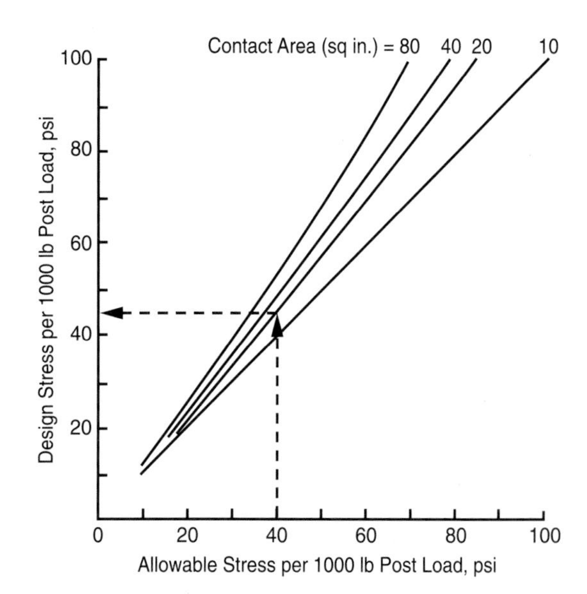

Fig. 5-12b. Design stress depends on allowable stress and contact area for the case of tensile stress at the slab top for slab edge post loads.

stresses) and in Fig. 5-12b for Fig. 5-11d for top surface stresses. For example, Fig. 5-12a is entered from the x-axis with an allowable stress of 40 psi. Following a vertical line up to the contact area line (in this case 20 in.^{2}), a design working stress of about 49 psi (per 1000 lb. post load) is determined from the y-axis.

For very heavy post loads, the required thickness of plain concrete slabs may be great enough that alternate designs should be considered, such as:

- integral or separate footings under each post or line of posts (post locations would have to be permanently fixed)
- structurally reinforced slabs with steel designed to take the tensile stresses
- use of a cement-treated subbase under the concrete slab
- pier or pile foundation for posts if there is potential for long-term slab settlement attributable to soil consolidation

Chapter 2 discusses the design of subbases and also the benefits they can provide to a concrete floor, including: better support; improved load transfer across the joint (especially for cement-treated subbases under a concrete floor slab subjected to very high loading conditions); a stable working platform; and reduced floor thickness.

DISTRIBUTED LOADS

Distributed loads act over large areas of a floor. The loads are primarily the result of material being placed directly on the floor in storage bays. For most plant and warehouse buildings, concentrated loads control floor design since they produce higher flexural stresses than distributed loads. However, after an adequate slab thickness has been selected to support the heaviest vehicle and post loads, the effects of distributed loads should also be examined. An example of distributed loading is shown in Fig. 5-13, with a graphic representation given in Fig. 5-14.

Two criteria control floor design where distributed loads are concerned:

- to prevent cracks in the aisleways due to excessive negative moment (tension in top of slab)
- to avoid objectionable settlement of the slab

Distributed loads can result in cracking in an unjointed aisle. Usually, distributed loads placed on floors are not great enough to cause excessive settlement of properly prepared and compacted subgrades. Although making the slab thicker can control cracking, it does not prevent slab settlement. For very heavy distributed loads on compressible subgrades, the possibility of soil consolidation should be examined by soil foundation engineering techniques.

Heavy distributed loads on portions of the slab may cause differential settlement and slab deformation. The slab bending stresses due to slab deformations may be additive to the negative bending stresses in aisles. Design of slabs subjected to differential slab settlement is beyond the scope of this publication. It is helpful to consult a foundation engineer in these cases.

Fig. 5-13. Distributed loads on a floor surface. (69656)

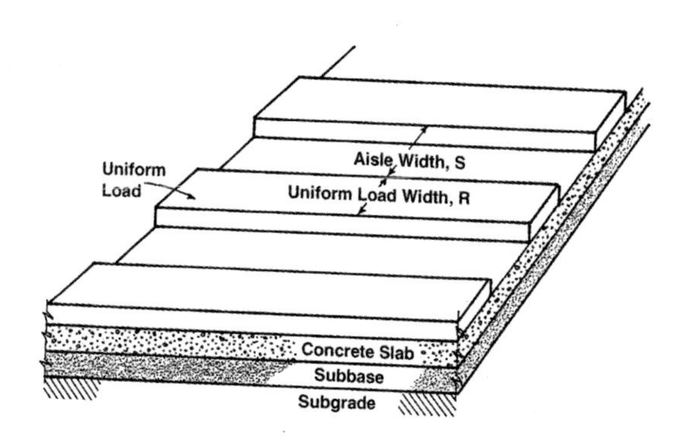

Fig. 5-14. Uniform or "strip" loading on a concrete floor.

Maximum Loads and Critical Aisle Width

For a given slab thickness and subgrade strength there is a critical aisle width for which the slab stress in the aisleway is maximum. As shown in Table 5-4, the allowable load for the critical aisle width is less than for any other aisle width. This means that both narrower and wider aisles will allow heavier loads to be placed on the slab. The critical aisle width exists when the maximum bending moment in the aisle, due to a load on one side of the aisle, coincides with the point of maximum moment, due to the load on the other side of the aisle. This doubles the negative bending moment (tension in top of slab) at aisle centerline. For other than the critical aisle width, the maximum bending moments, due to loads on each side of the aisle, do not coincide; the load on one side of the aisle may even counteract stress caused by a load on the other side.

Allowable Loads to Prevent Cracking in an Unjointed Aisleway

In an unjointed aisleway between distributed load areas, the maximum negative bending moment in the slab may be up to twice as great as the moment in the slab beneath the loaded area. As a result, one design objective is to limit maximum negative moment stresses in the aisleway so that a crack will not occur.

This publication presents two methods to determine the magnitude of allowable distributed loads based on this design objective: one uses tables, and the other involves figures. The use of the tables is discussed first.

Table 5-3 is used if the aisle and storage layout is variable and may be changed during the service life of the floor. If the layout is permanently fixed, Table 5-4 is used. In both of these tables, flexural stresses were computed based on the work of Hetenyi (1946). Rice used a similar approach, but because he made a large allowance for slab shrinkage stresses and considered zero uplift of joints in the aisleway as one of the design criteria, the allowable distributed loads are different (Rice 1957). As explained earlier in this chapter, shrinkage stresses may be disregarded in the design procedure. Also, it can be shown that loads up to the magnitudes shown in these tables produce very small joint uplifts that would be tolerable for most requirements. Slab stresses due to wheel loads at uplifted joints are less than the stresses in slabs not uplifted at the joint (Childs 1964).

It should be noted that the *k*-value of the subgrade, rather than the *k* on top of the subbase (if there is one), is used in Tables 5-3 and 5-4. This is appropriate for distributed loads covering large areas, while the use of the *k*-value on the top of the subbase is appropriate for concentrated loads.

Table 5-3. Allowable Distributed Loads, Unjointed Aisle (Nonuniform Loading, Variable Layout)

Slab thickness, in.	Subgrade *k*,[1] pci	Allowable load, psf[2]			
		Concrete flexural strength, psi			
		550	600	650	700
5	50	535	585	635	685
	100	760	830	900	965
	200	1,075	1,175	1,270	1,370
6	50	585	640	695	750
	100	830	905	980	1,055
	200	1,175	1,280	1,390	1,495
8	50	680	740	800	865
	100	960	1,045	1,135	1,220
	200	1,355	1,480	1,603	1,725
10	50	760	830	895	965
	100	1,070	1,170	1,265	1,365
	200	1,515	1,655	1,790	1,930
12	50	830	905	980	1,055
	100	1,175	1,280	1,390	1,495
	200	1,660	1,810	1,965	2,115
14	50	895	980	1,060	1,140
	100	1,270	1,385	1,500	1,615
	200	1,795	1,960	2,120	2,285

[1] *k* of subgrade; disregard increase in *k* due to subbase.
[2] For allowable stress equal to 1/2 flexural strength.
Based on aisle and load widths giving maximum stress.

Variable Storage Layout

Flexural stresses and deflections due to distributed loads vary with slab thickness and subgrade strength. They also depend on aisle width, width of loaded area, load magnitude, and whether or not there are joints or cracks in the aisleway. These additional variables are not always constant or predictable during the service life of a floor. Therefore, the allowable loads shown in Table 5-3, representing the most critical conditions, are suggested for practical design use where the aisle and storage layout is not predictable or permanent.

Since the allowable loads in Table 5-3 are based on the most critical conditions, there are no restrictions on the load layout configuration or the uniformity of loading. Loads up to these magnitudes may be placed non-uniformly in any configuration and changed during the service life of the floor. (Heavier loads may be allowed, as shown in Table 5-4, under restricted conditions of load configuration.)

The allowable loads in Table 5-3 are based on a safety factor of 2.0 (allowable working stress equal to one-half of the concrete's flexural strength). This is conservative. If the designer wants to incorporate other safety factors, the allowable working stress is first computed by dividing the

$$W = 0.123 \bullet f_t \sqrt{h \bullet k}$$

Table 5-4. Allowable Distributed Loads, Unjointed Aisle (Uniform Load, Fixed Layout)

Slab thickness, in.	Working stress, psi	Critical aisle width, ft**	Allowable load, psf					
			At critical aisle width	At other aisle widths				
				6-ft aisle	8-ft aisle	10-ft aisle	12-ft aisle	14-ft aisle
Subgrade $k = 50$ pci*								
5	300	5.6	610	615	670	815	1,050	1,215
	350	5.6	710	715	785	950	1,225	1,420
	400	5.6	815	820	895	1,085	1,400	1,620
6	300	6.4	670	675	695	780	945	1,175
	350	6.4	785	785	810	910	1,100	1,370
	400	6.4	895	895	925	1,040	1,260	1,570
8	300	8.0	770	800	770	800	880	1,010
	350	8.0	900	9.5	900	935	1,025	1,180
	400	8.0	1,025	1,070	1,025	1,065	1,175	1,350
10	300	9.4	845	930	855	950	885	960
	350	9.4	985	1,085	1,000	990	1,035	1,120
	400	9.4	1,130	1,240	1,145	1,135	1,185	1,285
12	300	10.8	915	1,065	955	915	925	965
	350	10.8	1,065	1,240	1,115	1,070	1,080	1,125
	400	10.8	1,220	1,420	1,270	1,220	1,230	1,290
14	300	12.1	980	1,225	1,070	1,000	980	995
	350	12.1	1,145	1,430	1,245	1,170	1,145	1,160
	400	12.1	1,310	1,630	1,425	1,335	1,310	1,330
Subgrade $k = 100$ pci*								
5	300	4.7	865	900	1,090	1,470	1,745	1,810
	350	4.7	1,010	1,050	1,270	1,715	2,035	2,115
	400	4.7	1,115	1,200	1,455	1,955	2,325	2,415
6	300	5.4	950	955	1,065	1,320	1,700	1,925
	350	5.4	1,105	1,115	1,245	1,540	1,985	2,245
	400	5.4	1,265	1,275	1,420	1,760	2,270	2,565
8	300	6.7	1,095	1,105	1,120	1,240	1,465	1,815
	350	6.7	1,280	1,285	1,305	1,445	1,705	2,120
	400	6.7	1,460	1,470	1,495	1,650	1,950	2,420
10	300	7.9	1,215	1,265	1,215	1,270	1,395	1,610
	350	7.9	1,420	1,475	1,420	1,480	1,630	1,880
	400	7.9	1,625	1,645	1,625	1,690	1,860	2,150
12	300	9.1	1,320	1,425	1,325	1,330	1,400	1,535
	350	9.1	1,540	1,665	1,545	1,550	1,635	1,880
	400	9.1	1,755	1,900	1,770	1,770	1,865	2,050
14	300	10.2	1,405	1,590	1,445	1,405	1,435	1,525
	350	10.2	1,640	1,855	1,685	1,640	1,675	1,775
	400	10.2	1,875	2,120	1,925	1,875	1,915	2,030
Subgrade $k = 200$ pci*								
5	300	4.0	1,225	1,400	1,930	2,450	2,565	2,520
	350	4.0	1,425	1,630	2,255	2,860	2,990	2,940
	400	4.0	1,630	1,865	2,575	3,270	3,420	3,360
6	300	4.5	1,340	1,415	1,755	2,395	2,740	2,810
	350	4.5	1,565	1,650	2,050	2,800	3,200	3,275
	400	4.5	1,785	1,890	2,345	3,190	3,655	3,745
8	300	5.6	1,550	1,550	1,695	2,045	2,635	3,070
	350	5.6	1,810	1,810	1,980	2,385	3,075	3,580
	400	5.6	2,065	2,070	2,615	2,730	3,515	4,095
10	300	6.6	1,730	1,745	1,775	1,965	2,330	2,895
	350	6.6	2,020	2,035	2,070	2,290	2,715	3,300
	400	6.6	2,310	2,325	2,365	2,620	3,105	3,860
12	300	7.6	1,890	1,945	1,895	1,995	2,230	2,610
	350	7.6	2,205	2,270	2,210	2,330	2,600	3,045
	400	7.6	2,520	2,595	2,525	2,660	2,972	3,480
14	300	8.6	2,025	2,150	2,030	2,065	2,210	2,480
	350	8.6	2,360	2,510	2,365	2,405	2,580	2,890
	400	8.6	2,700	2,870	2,705	2,750	2,950	3,305

* k of subgrade; disregard increase in k due to subbase.
** Critical aisle width equals 2.209 times radius of relative stiffness.
Assumed load width = 300 in.; allowable load varies only slightly for other load widths.
Allowable stress = one-half flexural strength (safety factor = 2.0).

28-day flexural strength, *MR*, by the safety factor; then the allowable load may be computed as:

where W = allowable load, psf
f_t = allowable working stress, psi
h = slab thickness, in.
k = subgrade modulus, pci

Fixed Storage Layout

As discussed in the previous section, slab stresses under distributed loads vary with aisle width, load width, load magnitude, and joint location. In a storage area where this layout is known and will remain fixed throughout the service life of the floor, the heavier distributed loads shown in Table 5-4 may be allowed (see box, Maximum Loads and Critical Aisle Width). These loads are based on limiting the negative moment in an unjointed aisleway so that an aisle crack does not occur.

Allowable Loads to Prevent Slab Settlement

In the previous discussion, the allowable loads were based on preventing a crack in an *unjointed* aisleway between loaded storage areas. If these cracks can be tolerated or if there are joints in the aisleways, there need be no such restriction on the load magnitude. In this case, the limit of load depends on the tolerable settlement of the slab.

Strip Loads

The general loading of long strips of storage areas separated by aisles is depicted in Fig. 5-14. Table 5-4 is based on a load width of 300 in. Because the allowable load varies only slightly for other load widths, this table can be used for a wide range of situations. However, for narrower strip loads (40, 60, and 100 in. wide), there are charts available to determine the allowable uniform load. Figs. 5-15a through 5-15e are not currently available in metric. Metric values should be converted to pounds and inches before using these charts, then the answer can be converted back to metric.

The charts consider the same aisle widths shown in Table 5-4, namely 6 ft to 14 ft in 2-ft increments. The load data, load width, and aisle width should be known. A safety factor of 2.0 can be used. The working stress is calculated by dividing the MR by the safety factor. One of Figs. 5-15a through 5-15e is chosen based on the appropriate aisle width, and a slab thickness is assumed.

For example, using Fig. 5-15a, the chart is entered at the right-hand side using an assumed slab thickness, read across horizontally to the modulus of subgrade reaction curve, then vertically up to the load width, then horizontally across to the stress. This is the first attempt in the iteration. This stress is per 1000 psf uniform load: if the value read from the chart is 200 psi and the load is 1500 psf, then the actual allowable stress would be 200 x (1500/1000) = 300 psi. If the chart-derived (allowable) stress is greater than the working stress, the slab thickness should be increased and the charts entered again until the stress per uniform load is less than the working stress. The optimum design occurs when the allowable stress matches the working stress.

The charts in Figs. 5-15a–e represent specific loading conditions at three chosen load widths to supplement the existing thickness design charts and tables. Each chart is for a uniform strip load at a single aisle width. This is slightly more specific than Table 5-4, which gives the maximum allowable load and critical aisle width for a single load width (300 in.). Table 5-4, however, gives critical load configurations adapted to almost any (fixed) loading condition. Table 5-3 places no restriction on where the load is placed or on the uniformity of loading. Computer programs (for analysis) would allow even greater design flexibility.

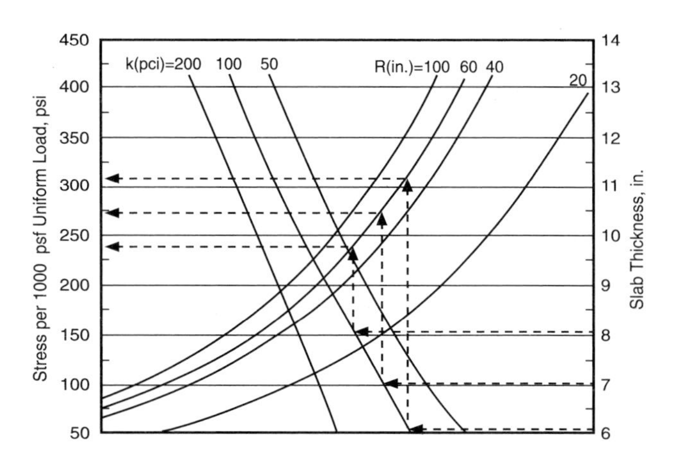

Fig. 5-15a. Design chart for uniform strip loading, aisle width of 6 ft.

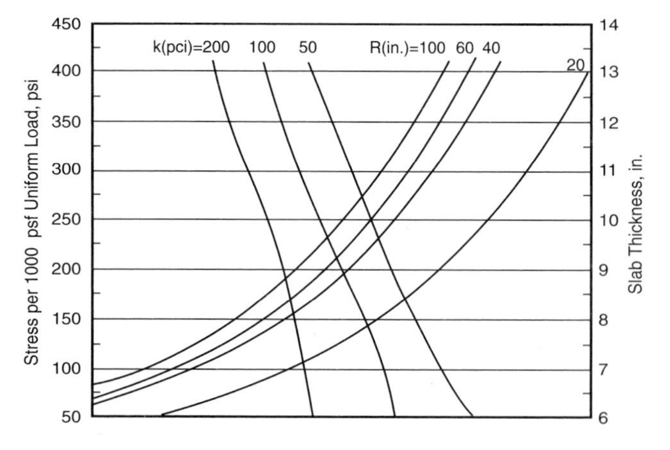

Fig. 5-15b. Design chart for uniform strip loading, aisle width of 8 ft.

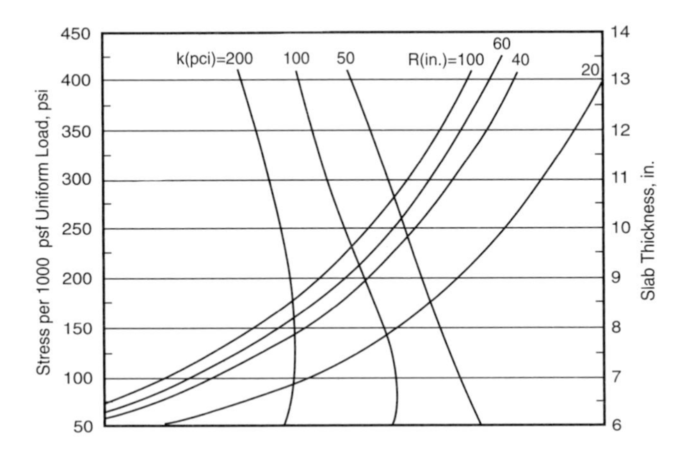

Fig. 5-15c. Design chart for uniform strip loading, aisle width of 10 ft.

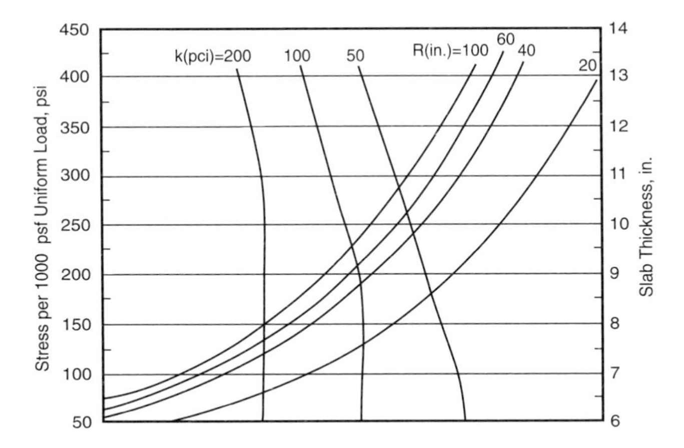

Fig. 5-15d. Design chart for uniform strip loading, aisle width of 12 ft.

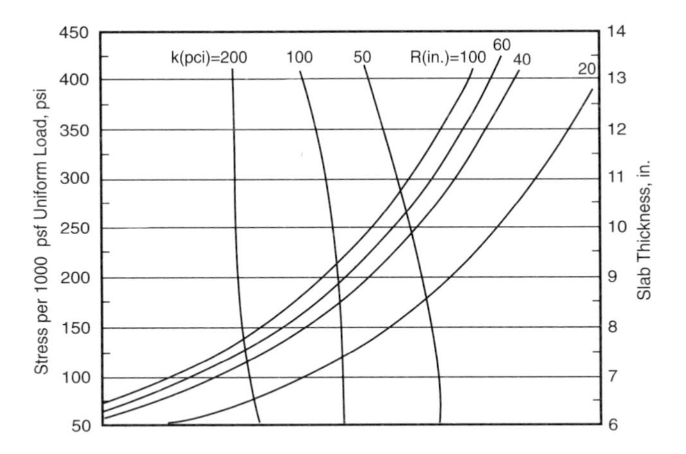

Fig. 5-15e. Design chart for uniform strip loading, aisle width of 14 ft.

Once the flexural stress criterion has been satisfied, the slab settlement should be checked. Slab settlement is caused by excessive pressures on the underlying soil. For concentrated loads, a greater slab thickness reduces the pressure on the soil. However, for distributed loads, slab thickness has virtually no effect on soil pressure—the soil pressure is equal to the distributed load plus the weight of the slab. Therefore, use of a thick slab does not reduce settlement under distributed loads.

The minimum settlement that may occur can be estimated by computing the elastic deflection using the subgrade modulus, *k* (Hetenyi 1946). A deflection profile should be computed for the entire slab width. This provides an estimate of the potential slab uplift in unloaded areas as well as downward deflection of the part of the slab that is directly under the load. These computed elastic deflections may be a fair estimate only if the soil is relatively incompressible.

For slabs on compressible subgrades, settlement under distributed loads may be considerably greater than the computed elastic deflection. If the distributed loads are heavy, the settlement should be estimated by methods used in soils engineering for spread or raft foundations. As described in most texts on soils engineering, data from test footings of different sizes are helpful in estimating the anticipated amount of settlement on compressible soils.

The tolerable settlement for floors depends on the operating requirement of the floor. Settlement may be considerably less than that allowed for foundations. For example, if a joint or crack exists in the aisle or within certain distances of the edge of the distributed load, the differential settlement between the loaded and unloaded sides of the joint or crack may cause a bump that is objectionable to wheeled traffic.

UNUSUAL LOADS AND OTHER CONSIDERATIONS

Special load configurations—unusual wheel or post configurations, tracked vehicles, vehicles with closely spaced axles or with more than 4 wheels per axle, very large wheel contact areas, strip loads—that differ substantially from those indicated in the previous discussions may be analyzed by one of the following methods:

- For concentrated loads (wheel or post loads), slab stresses, deflections, and soil pressures may be determined with influence charts or a computer program (ACPA 1992 and Pickett and Ray 1951). The influence charts may also be used for strip loads if the length of the load contact area is not great (see box, What Is a Large Contact Area?).
- When the dimensions of a load contact area are large (distributed loads in a storage bay or a strip load), the situation can be considered as a one-dimensional problem and the method of Hetenyi or finite element analysis computer program may be used (Hetenyi 1946).

What Is a Large Contact Area?

Influence charts cover a dimension of either 4 ℓ or 6 ℓ (see below), depending on the particular chart used.

The analysis of Hetenyi, as used for slab design, applies to loads of finite width and infinite length; however, the error is not significant if the length exceeds 6 ℓ where ℓ is defined by:

$$\ell = 4\sqrt{\frac{Eh^3}{12\left(1-\mu^2\right)k}}$$

where ℓ = radius of relative stiffness, in.
E = concrete modulus of elasticity, psi
h = slab thickness, in.
μ = Poisson's ratio
k = modulus of subgrade reaction, pci

To use this equation for metric measurements, convert the values to inch-pound quantities, do the calculations, then convert back to metric.

In either case, the controlling design considerations will be similar to those indicated in Fig. 5-1 based on the size of the load contact areas. (For strip loads, the controlling design considerations cannot be expressed in terms of contact area; it may be necessary to analyze several of the design considerations indicated in Fig. 5-1 to determine which one is most critical.)

A wide range of design situations can be analyzed with numerous finite element computer programs available for stress and deflection analysis of foundation mats, pavements, and slabs on grade (PCA 1995c).

Vibrating loads, such as those caused by heavy generators or compressors, may require a special foundation design and are beyond the scope of this discussion.

Support Losses Due to Erosion

The thickness design method in this publication focuses on keeping the flexural stress within safe limits to prevent fatigue cracking of concrete. The method assumes that the slab maintains contact with the ground underneath it. But if the slab loses support from the subgrade, the calculated stress can increase from 5% to 15%, which could affect the thickness design (Wu and Okamoto 1992). The lack of support increases slab stresses and strains under load and can lead to premature cracking that is not fatigue related.

Because slab failure is a potential outcome, the loss of support deserves consideration. Such information was developed from concrete highway and street pavement research. Climatic erosion conditions that lead to loss of support include:

- voids under the slab
- joint faulting
- pumping
- loss of shoulder material

While some of these effects are related primarily to highways and pavements, some could become considerations on interior slabs. Fortunately, a floor slab is subjected to a much milder range of temperature, humidity, and weather than a pavement. Two important differences between pavements and floors are (1) the quantity and frequency of water exposure (rainwater), and (2) the loads on floors are not imposed by high-speed vehicles, though the load magnitude may be quite large. If it is suspected that the slab has lost contact with the subgrade/subbase, the effect on the slab stresses should be investigated using established equations (Wu and Okamoto 1992).

CHAPTER 6

JOINTS, REINFORCEMENT, AND CRACK CONTROL

Cracks in concrete floors are often caused by restrained volume changes that create tensile stresses. When tensile stresses exceed concrete tensile strength, the concrete cracks. The possibility of random cracking exists because cooling contraction and drying shrinkage are unavoidable, inherent properties of hardened concrete.

Random cracking in concrete should be controlled, and there are many effective ways to accomplish this. One primary consideration is to minimize volume changes of the hardened concrete as discussed in Chapter 4. Other ways include jointing, use of reinforcing bars, or welded wire fabric. Fibers can help control plastic shrinkage cracks. Post-tensioning or shrinkage-compensating concrete can also be used, although these methods add a premium to the cost of the floor and require contractors with extensive experience and specialized equipment and materials. Post-tensioning and shrinkage-compensating concrete are described briefly later in this chapter, but are otherwise beyond the scope of this book.

Joints permit concrete to move slightly, thus reducing restraint and relieving stresses that cause random cracking. While joints are aesthetically more pleasing in appearance than random cracks, they may require filling and later maintenance to control spalling at the edges.

A jointing plan, or joint layout, should be prepared by the floor designer for all slab-on-grade work. Stating the type, number, and location of floor joints allows for better cost estimates and fewer mistakes during construction. (See Joint Layout later in this chapter for more information.)

JOINT CLASSIFICATION BY FUNCTION

Isolation Joints

Isolation joints permit horizontal and vertical movement between the abutting faces of a floor slab and fixed parts of the building, such as walls, columns, and machinery bases. Isolation joints completely separate the floor from the adjoining concrete, allowing each part to move independently without damage to the other. Columns on separate footings are isolated from the floor slab either with a circular- or square-shaped isolation joint. The square shape should be rotated to align its corners with control and construction joints, as shown in Fig. 6-1. The joint material should extend through the entire slab depth and be thick enough to allow for some compressibility or other movement (see Fig. 6-2).

Contraction Joints

Contraction joints (also called control joints) relieve stresses caused by restrained curling, cooling contraction, or drying shrinkage. With proper spacing and timely installation, these joints help to control random cracking by allowing horizontal slab movement. [Note: HIPERPAV, software

Fig. 6-1. Box-out for isolation joint. Isolation joints can be circular or square shaped and are used between the floor slab and fixed parts of building, such as walls, columns, and machinery bases. (69611)

Fig. 6-2. Joint material can be wrapped around circular members or can be attached to formwork before concreting begins. (69612)

sponsored by the FHWA, can be used to evaluate cracking potential.]

Contraction joints can be made in several ways:

- sawing a continuous straight slot in semi-hardened (slightly soft) concrete with special early-cut saws
- sawing a continuous straight slot in the hardened concrete
- hand grooving fresh concrete during finishing (not practical for floors more than 100 mm [4 in.] thick)
- installing premolded plastic or metal inserts during placing and finishing

Whether joints are made by hand grooving, inserting premolded materials, or sawing slightly soft or hardened concrete, they should usually extend into the slab to a depth of one-fourth the slab thickness (see Figs. 6-3, 6-4, and 6-5). The objective is to form a plane of weakness in the slab so that the crack will occur along that line and nowhere else, as shown in Fig. 6-3.

When floors carry vehicular traffic and joint fillers are needed, saw cutting the joint is the only acceptable method. Hand-grooved joints have rounded edges that are undesirable. Premolded inserts should be avoided because they may be disturbed during finishing. If the inserts are not plumb, spalling of the hardened concrete at joint edges is likely.

Joints in industrial and commercial floors are usually cut with a saw. Timing of joint sawing is critical. To minimize tensile stresses and random cracking caused by restrained curling or cooling contraction, sawing should be done before concrete cools appreciably. Slabs are especially vulnerable to curling and contraction stresses within the first 6 to 18 hours after concrete placement, when tensile strength is very low. Joints should be sawn as soon as the concrete is hard enough that the sawing does not ravel joint edges or dislodge coarse aggregate particles.

Lightweight, high-speed, early-cut saws have been developed to permit joint sawing very soon after floor finishing, sometimes within 0 to 2 hours (see Fig. 6-6). At this point, the concrete is very weak, having a strength of about 1 MPa to 3.5 MPa (150 psi to 500 psi). By cutting the slab before stress builds up, the mechanics of slab cracking that relate slab thickness to depth of cut are changed. In other words, if the cut is sawn within a few hours after final finishing, random cracking can be controlled even if the cut is less than one-fourth of the slab depth (PCA 1995b).

Most early-cut saws have a small-diameter diamond saw blade that extends through a slot in a metal skid plate and cuts a narrow slot about 20 mm to 25 mm (3/4 in. to 1 in.) deep. The largest early-cut saws can make cuts up to 75 mm (3 in.) deep, which the saw manufacturer recommends for slabs up to 660 mm (26 in.) thick. Some contractors use the lightweight saw for early crack control, then chase the same saw cut with a larger blade conventional saw 12 to 24 hours after concrete placement. This second saw pass increases the sawcut depth to at least one-fourth the slab depth.

In floors with welded wire fabric or reinforcing bars, there are different ways of handling steel at the joints, depending on how the joint should function (Ringo 1991):

- open/working joint—cut the steel (provide dowels if load transfer is needed)
- closed joint—continue steel through the joint (this can lead to cracking within the panel)

Designers may choose to partially continue steel through the joint in some instances. In any case, it is important to realize that how the steel is designed will affect the operation of the joint and the potential for cracking within slab panels (private communication with C. Bimel).

Fig. 6-3. Contraction joint formed by a saw cut. This side shot of a slab shows a joint extending about one-fourth of the slab depth, which usually ensures that the crack will form beneath the joint. (4434)

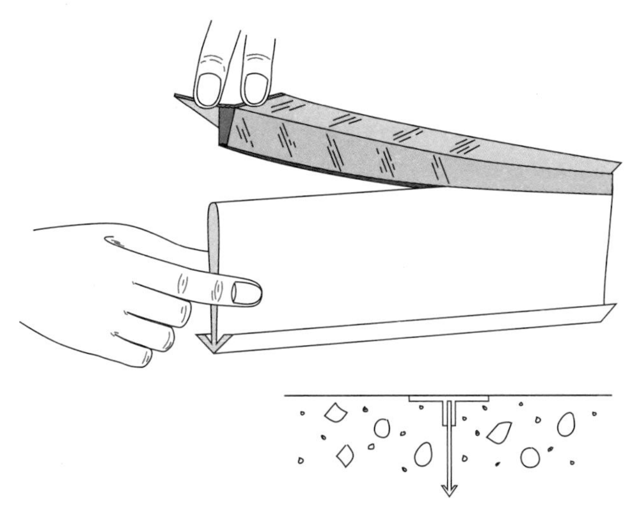

Fig. 6-4. A contraction joint can be formed by a plastic insert, but it is difficult to hold the insert vertical during concrete placement.

Fig. 6-5. Contraction joint formed by a hand grooving tool. (69614)

Fig. 6-6a,b. Early cut saws permit fast sawing of joints and reduce the occurrence of random cracking. (69615, 69657)

Construction Joints

Construction joints (Fig. 6-7) are stopping places and form the edge of each day's work. They are frequently detailed and built to function as and align with contraction joints or isolation joints. For many installations, construction joints are installed along some of the column lines, while contraction joints are sawn at other column lines and between joints at column lines.

Whenever continuous concrete placement will be interrupted for 30 minutes or more, a *bonded or tied construction joint* should be formed and deformed tie bars added (see Stoppage of Work later in this chapter). For floors exposed to small, hard-wheeled traffic, construction joints should be finished tight (no radius). Any reinforcement in the slab should be continuous through a bonded construction joint unless the joint is functioning as a contraction or isolation joint. It is common practice to return later to bonded construction joints and sawcut them to a depth of 25 mm (1 in.) to create a reservoir for joint fillers. After filling, both the traffic-bearing ability and appearance of the joint are improved.

Fig. 6-7. When concrete for the adjacent slab is placed, this construction joint will form a butt joint with the new slab. (69616)

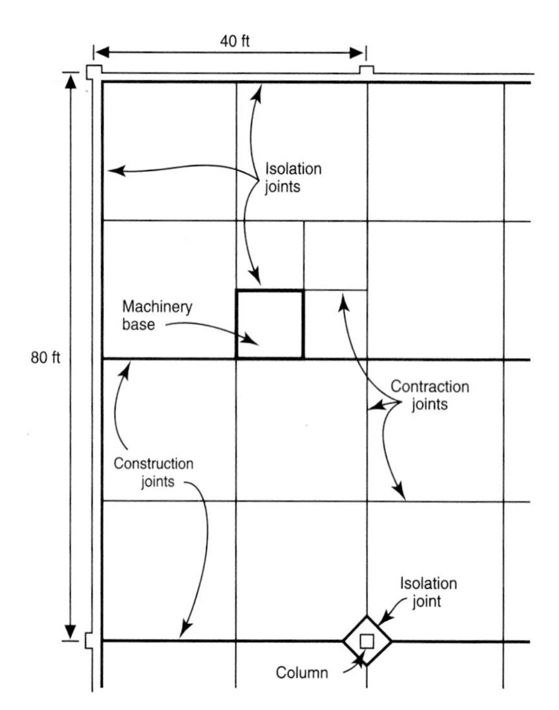

Fig. 6-8. This sample joint layout indicates typical locations for the various joint types.

JOINT LAYOUT

A sample joint layout is illustrated in Fig. 6-8. All three types of joints—isolation, contraction, and construction—will likely be used in a slab installation. Isolation joints are provided around the perimeter of the floor where it abuts the walls and around all fixed elements that restrain movement of the slab in a horizontal plane. This includes columns and machinery bases that pass through the floor slab. With the slab isolated from other building elements, the remaining task is to locate and correctly space contraction joints to eliminate random cracking. Proper planning of construction joints allows them to coincide with contraction joint placement. This allows breaking the job into portions that can be built in a single day, and yet minimizes the number of joints in the floor.

JOINT SPACING

Closely spaced joints help to reduce the number of random cracks between joints. They also decrease the amount of uplift if curling occurs at the joint. The most economical floor construction is a plain concrete slab (no reinforcement) with short joint spacing supported by a uniform subgrade.

For unreinforced concrete slabs, the maximum recommended spacing between joints depends primarily on:

- slab thickness
- cooling contraction and drying shrinkage potential
- curing environment

Contraction joint spacings recommended in Tables 6-1a and 6-1b will minimize random cracking. Experience with local materials and job practices may indicate that greater spacings than those shown in the table can be used without causing extensive cracking within floor panels.

Joints can usually be spaced further apart in thicker slabs. A rule of thumb for plain slabs constructed with 100-mm to 150-mm (4-in. to 6-in.) slump concrete is that joint spacing should not exceed 24 slab thicknesses for concrete made with less than 19-mm (3/4-in.) top-size coarse aggregate. For the same slump, but with concrete containing 19-mm (3/4-in.) or larger coarse aggregate, joint spacing should not exceed 30 slab thicknesses. When concrete with less than a 100-mm (4-in.) slump is used, suitable joint spacing should not exceed 36 slab thicknesses, provided that the slump reduction is due to the reduction of the amount of concrete mix water, which is the significant factor in reducing concrete drying shrinkage.

When joint spacing is increased, it is common to use armored edges at construction joints. Armored joints (steel angle or steel plate) are generally sealed with an elastomeric sealant.

Joints should be spaced closer together if concrete cools significantly at an early age. Early cracking caused by cooling may occur if the difference between first-night surface temperature and maximum concrete temperature after placement exceeds 6°C to 8°C (10°F to 15°F). Later cracking caused by drying shrinkage may occur if high-shrinkage concrete is used or if the concrete is improperly cured in a dry environment. Even at the suggested spacing, extensive cracking may occur if the concrete has a high-shrinkage potential, if it is improperly cured, or if it is exposed to early-age temperature drops mentioned above. The recommended spacings should be appropriate for the majority of floor construction projects, as they are based on jobsite conditions that are neither ideal nor poor, but somewhere between those extremes.

Table 6-1a. Spacing of Contraction Joints in Meters*

Slab thickness, mm	Maximum-size aggregate less than 19 mm	Maximum-size aggregate 19 mm and larger
125	3.0	3.75
150	3.75	4.5
175	4.25	5.25**
200	5.0**	6.0**
225	5.5**	6.75**
250	6.0**	7.5**

* Spacings are appropriate for slump between 100 mm and 150 mm. If concrete cools at an early age, shorter spacings may be needed to control random cracking. (A temperature difference of only 6°C may be critical.) For slump less than 100 mm, joint spacing can be increased by 20%.

** When spacings exceed 4.5 m, load transfer by aggregate interlock decreases markedly.

Table 6-1b. Spacing of Contraction Joints in Feet*

Slab thickness, in.	Maximum-size aggregate less than 3/4 in.	Maximum-size aggregate 3/4 in. and larger
5	10	13
6	12	15
7	14	18**
8	16**	20**
9	18**	23**
10	20**	25**

* Spacings are appropriate for slump between 4 in. and 6 in. If concrete cools at an early age, shorter spacings may be needed to control random cracking. (A temperature difference of only 10°F may be critical.) For slump less than 4 in., joint spacing can be increased by 20%.

** When spacings exceed 15 ft, load transfer by aggregate interlock decreases markedly.

While a given joint spacing may control random cracking within the panel, that same spacing may not adequately limit the width of the opening at the joint. This may affect load transfer across the joint. Crack width (below contraction joint sawcuts or formed inserts) depends on the amount of cooling contraction and drying shrinkage. Contraction joint crack widths can be calculated as shown in the following example:

Metric

Joint spacing = 4575 mm = (4.575 m)

Concrete cooling = 11.1°C from initial hardening to facility condition in winter

Temperature coefficients = 9×10^{-6}/°C

Drying skrinkage coefficient = 100×10^{-6}

Cooling contraction = $11.1 \times 9 \times 10^{-6} \times 4575$ mm = 0.46 mm

Drying shrinkage = $100 \times 10^{-6} \times 4575$ mm = 0.46 mm

Total crack width = <u>0.92 mm</u>

In./lb.

Joint spacing = 15 ft = 180 in.

Concrete cooling = 20°F from initial hardening to facility condition in winter

Temperature coefficients = 5×10^{-6}/°F

Drying shrinkage coefficient = 100×10^{-6}

Cooling contraction = $20 \times 5 \times 10^{-6} \times 180$ in. = 0.0180 in.

Drying shrinkage = $100 \times 10^{-6} \times 180$ = 0.0180 in.

Total crack width = <u>0.036 in.</u>

Note that this example shows a crack width that is slightly exceeding the aggregate interlock effectiveness limit (for slabs up to 175 mm [7 in.] thick). Concrete with a higher shrinkage potential or an increase in the temperature difference (from installation to operation) will push the crack beyond the limit.

For long joint spacings or for heavily loaded slabs, round, smooth, steel dowel bars can be used as load transfer devices at contraction or construction joints (see Load Transfer by Dowels, page 56). However, dowels are not needed when slab edge loading is considered in the design process and/or thickened edges are used.

LOAD TRANSFER AT JOINTS

When floor thickness design is based on interior-of-slab loading, load transfer is needed at joints. Load transfer across closely spaced contraction joints (less than 4.5 m [15 ft]) is provided by the interlocking action of the aggregate particles at the fracture faces of the crack that forms below saw cuts or formed notches (Colley and Humphrey 1967). This is called aggregate interlock. Load transfer across wider cracks is developed by a combination of aggregate interlock and mechanical devices such as dowel bars. When load transfer is effective, the stresses and deflections in the slab near the joint are low, and forklift and industrial trucks move smoothly across the joint without damaging it, the vehicle, or the cargo.

Load Transfer by Aggregate Interlock

Aggregate interlock at joints in concrete pavements was investigated by the Portland Cement Association (Colley and Humphrey 1967, Nowlen 1968). Joint effectiveness measures the ability (percentage) of the joint to transfer load from the loaded slab to the unloaded slab (Fig. 6-9). If load transfer at a joint were perfect, deflections of the loaded and unloaded slabs would be equal and the effectiveness in terms of slab edge deflections would be 100%. If there were no load transfer at a joint, only the loaded slab would deflect and the effectiveness would be zero. Observations from the investigation are also applicable to industrial floors on ground.

Effectiveness of aggregate interlock load transfer depends on:

Fig. 6-9. Determining load transfer effectiveness involves loading one side of the joint and measuring slab deflection on both sides of the joint. (69658)

- joint opening
- slab thickness
- subgrade support
- load magnitude and number of repetitions
- aggregate angularity

Effect of Joint Opening and Slab Thickness

Load-transfer effectiveness was determined for joint openings ranging from 0.4 mm to 2.2 mm (0.015 in. to 0.085 in.). Effectiveness decreased as the joint openings became wider. An opening of 0.9 mm (0.035 in.) or less showed good load-transfer effectiveness and endurance. Joints with larger openings transferred load less effectively. Joint openings of 0.6 mm (0.025 in.) were almost 100% effective in terms of slab edge deflection.

Thick slabs with wide openings were found to be as effective as thin slabs with narrower openings. On the basis of edge deflections, 0.9 mm (0.035 in.) openings in 230-mm (9-in.) thick slabs gave the same 60% effectiveness as 0.6 mm (0.025 in.) openings for 180-mm (7-in.) thick slabs (Colley and Humphrey 1967).

Effect of Subgrade Support

Concrete floors on ground do not necessarily require strong subgrade support to carry design loads successfully. However, higher k-values do increase joint effectiveness. Slabs on silts and clays with k-values of 14 MPa/m to 27 MPa/m (50 pci to 100 pci) start losing aggregate interlock effectiveness after only a few repetitive loading cycles. Sandy soils with k = 54 MPa/m (200 pci) maintain 50% joint effectiveness (on the basis of slab edge deflections) through one million load cycles. Sand-gravel and cement-treated subbases with k = 81 MPa/m (300 pci) or higher, can keep aggregate interlock effectiveness at higher percentages (>50%) through one million load cycles.

Effect of Load

Aggregate interlock effectiveness decreases as the magnitude of repetitive loads increases. Joints that successfully provide load transfer for one load condition may break down under a new, heavier load condition. An accurate estimate of maximum expected load is essential to keep joints functioning properly in terms of load transfer.

Effect of Aggregate Angularity

Concrete made with crushed gravel or crushed stone provides more effective aggregate interlock and load transfer than concrete made with natural gravel. The higher effectiveness is attributed to improved interlock provided by angularity of the crushed coarse aggregate particles.

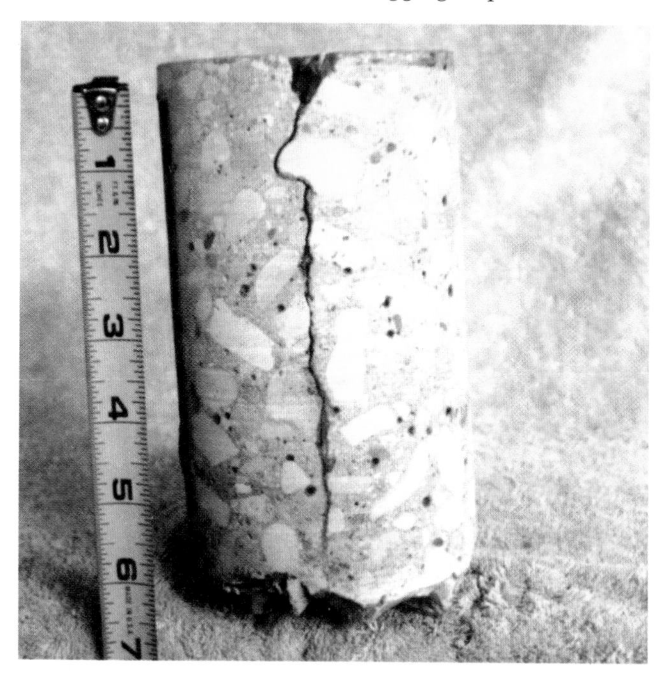

Fig. 6-10. Irregular faces of a crack that forms below a sawn control joint provide adequate aggregate interlock for load transfer when short joint spacings are used. (67189)

Load Transfer by Dowels

When joint openings become wider, load transfer by aggregate interlock is not as effective. Joint openings will be larger when longer joint spacings are used. To keep joint widths small, the spacing between successive joints should be kept small. For many industrial floors, however, contraction joint spacing is dictated by location of columns or racks within the structure. The cost of forming and filling joints or the owner's desire to limit the number of joints may also result in larger joint spacings. Dowels can provide load transfer at contraction and construction joints when larger joint spacings are used and for heavily loaded industrial floors. Dowels are usually smooth round steel bars. Recommended dowel sizes and spacings are shown in Tables 6-2a and 6-2b.

Table 6-2a. Dowel and Tiebar Sizes and Spacings (Metric)

	Dowels		
Slab depth, mm	Diameter, mm	Total length, mm	Spacing, mm, c to c
125	16	300	300
150	19	350	300
175	22	350	300
200	25	350	300
225	29	400	300
250	32	400	300
	Tiebars		
125	13	760	760
150	13	760	760
175	13	760	760
200	13	760	760
225	13	760	760
250	13	760	760

Table 6-2b. Dowel and Tiebar Sizes and Spacings (In.-Lb.)

	Dowels		
Slab depth, in.	Diameter, in.	Total length, in.	Spacing, in., c to c
5	5/8	12	12
6	3/4	14	12
7	7/8	14	12
8	1	14	12
9	1-1/8	16	12
10	1-1/4	16	12
	Tiebars		
5	#4	30	30
6	#4	30	30
7	#4	30	30
8	#4	30	30
9	#5	30	30
10	#5	30	30

Dowels resist shear and help to reduce deflections and stresses as loads cross the joint. At least one-half of each dowel bar should be coated with a bondbreaker or covered with a sleeve to prevent bond with concrete. Sleeves with the capped or closed end extending beyond the body of the dowel are used when slab expansions are anticipated after slab installation. This accommodates expansion or contraction of concrete at the joint and ensures free longitudinal movements. The dowels continue to function after the slabs shrink and pull apart. Though round bars are the most common dowel shape, other load transfer devices, such as square smooth bars and flat plates, have recently become available. Proponents of these proprietary systems claim they offer advantages over standard round bars.

The square bars are used with a special clip (sleeve) that covers one-half of the bar. The clip is made from a hard plastic, but its sides contain a compressible material. This design maintains the vertical alignment between the two slab faces, while allowing small lateral horizontal movements. An important benefit of this system is that it reduces cracking associated with misaligned bars and with normal slab movements at joints (Schrader 1999). Additionally, these square dowels transfer load better than round bars and control slab curling.

Another newer dowel shape is the flat plate (Walker and Holland 1998). Plate dowels are rectangular or diamond-shaped (square). Rectangular plates are used for contraction (control) joints, and diamond plates are used for construction joints. Compared to the traditional round smooth dowel bar, plate dowels:

- are easier to place
- allow horizontal slab movement
- minimize stress concentrations both on the slab and on the dowel
- are a more efficient use of material (steel for reinforcement)
- may be more cost-effective

Because plate dowels are relatively new, there are no officially recognized size and spacing recommendations, but there are tables to convert round smooth dowel bars to plate dowels (Walker and Holland 1998). Plate dowels should be spaced no more than 610 mm (24 in.) apart to limit slab deflection and are otherwise designed to give equivalent performance to traditional smooth round dowels, including:

- the same stiffness (vertical deflection between the two slabs joined by the dowel)
- the same bearing stress on concrete
- the same stresses in the dowel (flexural and shear)

Dowels, whether bars or plates, are installed at slab mid-depth, parallel to the floor surface and slab centerline. Round dowel bars not placed parallel will restrain joint movement and may thus cause random cracking. Dowel bar cages can be used to assure proper bar positioning in concrete floor construction. (This technology is borrowed from mainline paving projects.) The new dowel shapes—square bars and flat plates—may be more forgiving than traditional round bars in this respect. Both the plate geometry and the special clip for the square bars allow for horizontal movement, which can prevent unwanted cracking.

Dowels should not be confused with tie bars. Because dowels are smooth, they allow joints to open and close and are used at contraction joints and at construction joints that are designated to also allow slab end movement. Joints that contain tie bars, on the other hand, cannot function as contraction joints. Tie bars have raised bumps on their surface (they are deformed), so they bond to concrete on both sides of the joint and prevent joint and crack widening. This helps induce aggregate interlock and makes for better load transfer. This restraint to movement, however, can lead to random cracking. The one function that dowels and tie bars have in common is the ability to transfer vertical loads (shear) across joints. Tie bars also help to minimize slab edge upward warping (curling) deformations. This keeps floors flat across joints. Suggested tie bar sizes and spacings are given in Tables 6-2a and 6-2b.

Load Transfer by Keyways

Keyways have been used for load transfer at construction joints in some floors. Keyed joints are also known as tongue-and-groove joints. They consist of a projection on one joint face with a matching indentation on its abutting face. To construct this type of joint, a beveled strip of wood or a preformed key is attached to a form face. Concrete is cast, the form is stripped, and a depression (groove) remains in the edge of the slab. This groove is coated with bondbreaker, and when the adjacent slab is placed concrete fills the groove, creating a perfect fit projection (tongue). In theory, this keys the slabs together.

In practice, most keyed joints do not remain tight. As the floor slabs shorten due to shrinkage, the key loses contact with its matching recess (see Fig. 6-11). Then slab edges deflect as loads roll over the joint. This loss of load transfer is an inherent weakness of keyed joints, especially in heavily loaded floors. In addition, spalling occurs in the upper shoulder of the female side of the joint. Due to poor load transfer effectiveness and spalling, keyed joints should not be used in floors exposed to lift truck and other traffic.

Fig. 6-11. This core hole drilled over a keyed joint shows how the two faces have separated, rendering the keyway ineffective. (67193)

FILLING AND SEALING JOINTS

There are three options for treating joints: they can be filled, sealed, or left open. Joint filling is mandatory for all joints exposed to hard-wheeled traffic. In some cases where the traffic loading is lighter, joints can be sealed instead. The difference between a filler and sealer is the hardness of the material, as fillers are more rigid than sealers and provide support to slab edges, and thus minimize edge spalling. For commercial slabs and lightly used industrial floors, the joint may be left unfilled or unsealed.

Contraction and construction joints in floor areas exposed to solid rubber, hard urethane, or nylon casters and steel-wheel traffic should be filled with a semi-rigid epoxy or polyurea that provides lateral support to vertical edges of the joint sawcut. Fillers should not become hard or brittle in service. A filler for this purpose also must have low-range tensile strength and adhesion to concrete so that it will yield before the concrete does if the slab moves. In Section 9.10, ACI 302.1R recommends using epoxy or polyurea filler with 100% solids and a minimum Shore Hardness of A 80 when measured in accordance with ASTM D 2240. This material should be installed full depth in sawcut joints, *without* a backer rod, and *flush* with the floor surface.

For deeper cuts or where a crack extends below the joint, filler sand is brought to within 50 mm (2 in.) of the surface, and the joint filling is completed with semi-rigid filler smoothed flat to the floor surface.

Joint filling may be omitted if the floor is exposed only to foot traffic or low-pressure pneumatic tires. If radon is a concern, joints should be filled (see Chapter 2). Where there are wet conditions or concerns about hygiene or dust control, joints can be sealed with a flexible elastomeric sealant. A typical sealant application would be a polyurethane elastomer with a Shore Hardness of A 35 to A 50 installed 13 mm (1/2 in.) deep over a compressible backer rod.

Prior to filling joints formed by inserts, it is recommended to saw over the insert alignment to the full insert depth. Fig. 6-12 shows a contraction joint ready for filling. The filler reservoir should be sawn at least 25 mm (1 in.) deep. The sawing should also eliminate rounding near the surface edges.

Before filling any sawn joints, they must be cleaned to ensure good bond between the filler and bare concrete. All saw cuttings and construction debris should be removed. Vacuuming is preferred to blowing out the joints with compressed air.

Filling the cleaned reservoir space with a semi-rigid filler should be delayed as long as possible to permit the joint to open as slab shrinkage occurs; this promotes good joint filling and better joint filler performance. To ensure that the joint is flush with the floor surface, the joint is overfilled, allowed time for curing, then shaved or ground flat. Joints should always be filled before subjecting them to hard-wheel traffic.

Following filling or sealing, a joint that is moving could exceed the filler extensibility and lead to a separation at the joint wall (adhesion failure) or within the filler (cohesion failure). When filler separation occurs, the voids should be refilled with the same original filler or a companion material supplied by the same manufacturer. If the filler becomes loose to the touch, it should be totally removed and replaced. Separated sealants must be removed and replaced.

Isolation joints, which are designed to accommodate movement, can be filled by removing ("raking out") the top portion of the premolded compressible material, then filling the cavity with an elastomeric material. Alternately, a premolded joint former with a removable insert can be used to provide the reservoir for the sealant.

If joints do not function properly, there may be edge spalling at the joint face, or random cracks may appear in the slab. The greatest portion of floor repair and maintenance is for joint edge deterioration and crack correction. See Chapter 10 for information on repairing joints and cracks.

Fig. 6-12. Cut-away view (core) of a joint showing the reservoir that will accept a joint filler. (69659)

DISTRIBUTED STEEL FOR FLOORS ON GROUND

Table 6-3. Reinforcement for Floors on Ground

Is reinforcement necessary?	
NO	• With uniform support and short joint spacing
YES	• When long joint spacing is required • When joints are unacceptable in floor use

Distributed steel refers to welded wire fabric or reinforcing bars placed in concrete slabs. The relatively small amount of steel is used to hold together fracture faces when (random) cracks form. Floors that contain steel bars or fabric are referred to as reinforced slabs (see Fig. 6-13). When floor joints are spaced less than 4.5 m (15 ft) apart, distributed steel is not needed in concrete floors on ground unless cracks must be tightly held together. Normally, short panel lengths will control cracking between joints. The short panels will reduce the total shrinkage in any one panel to a small enough value that the contraction joint may be satisfactory to assure aggregate interlock.

In slabs with long joint spacings, the purpose of distributed steel is to hold random intermediate cracks tight. Floor designers should therefore be aware and should accept that random intermediate cracks may, can, and do occur on concrete slabs having long joint spacings and containing distributed steel.

Since critical flexural stresses occur in both the top and bottom of concrete floors, steel should be placed in two layers to best resist stresses. Though two layers of steel may be ideal for resisting stress, double layers of steel are not always economically justified, and there may not be enough space to accommodate two layers. Ease of placement, or "constructibility," should also be considered.

Distributed steel can minimize the number of joints required in floor slabs, especially when long joint spacings are selected. Cracks due to restrained contraction or curling may occur, but sufficient amounts of distributed steel will hold the cracks tightly closed to permit load transfer through aggregate interlock.

Distributed steel does not prevent cracking, compensate for poor subgrade preparation, or significantly increase load-carrying capacity of the floor. Distributed steel reinforcement, both wire mesh or bar mats, should be terminated (cut) within 50 mm (2 in.) of both sides of con-

Fig. 6-13. A reinforced slab containing plate dowels and a mat of rebar. (69617)

traction and construction joint alignments. Failure to cut all wires and bars would transfer shrinkage and contraction restraints across the joints and would contribute to intermediate random cracks and variable joint openings. This would make joint filling and sealing maintenance more difficult.

Subgrade Drag Formula for Determining Amount of Distributed Steel

The traditional subgrade-drag formula provides one method for determining the amount of distributed steel needed. The cross-sectional area of steel per lineal foot of slab width is given by:

$$A_s = \frac{FLw}{2f_s}$$

in which

A_s = cross-sectional area of steel, in square inches per lineal foot of slab width

F = coefficient of subgrade friction. (Designers use 1.5 or 2.0 for pavements; 1.5 is recommended for concrete floors on ground.)

L = slab length (or width if appropriate) between free ends, in feet. (A free end is any joint free to move in a horizontal plane.)

w = weight of slab, in pounds per square foot. (For normal-weight concrete, designers use 12.5 pounds per inch of floor thickness.)

f_s = allowable working stress of reinforcement, in pounds per square inch. (The working stress of steel is usually 0.67 or 0.75 times the yield strength of the steel in pounds per square inch.)

Whether welded-wire fabric, high-yield steel bars, or mild steel bars are used, the subgrade-drag formula often results in calculated steel areas less than 0.1% of the cross-sectional area of the slab. However, based on experience and slab performance surveys, some researchers, floor designers, and contractors suggest that for best performance, a minimum amount of distributed steel may be needed (CCA 1985, Gilbert 1992, Marias and Perrie 1993). In an FHWA study of jointed concrete pavement performance, pavement sections with more than 0.1% reinforcing steel exhibited less deteriorated transverse cracking. Sections with less than 0.1% reinforcing steel often displayed a significant amount of deteriorated transverse cracking, especially in colder climates. Because of this, the study recommended using a minimum of 0.1% reinforcing steel based on cross-sectional area, with larger amounts required for harsher climates and longer slabs.

Similar recommendations have been made for concrete floors. One approach for floors with long joint spacings is to use the subgrade-drag equation in conjunction with a minimum allowable reinforcement ratio. A minimum reinforcement ratio of 0.15% has been suggested

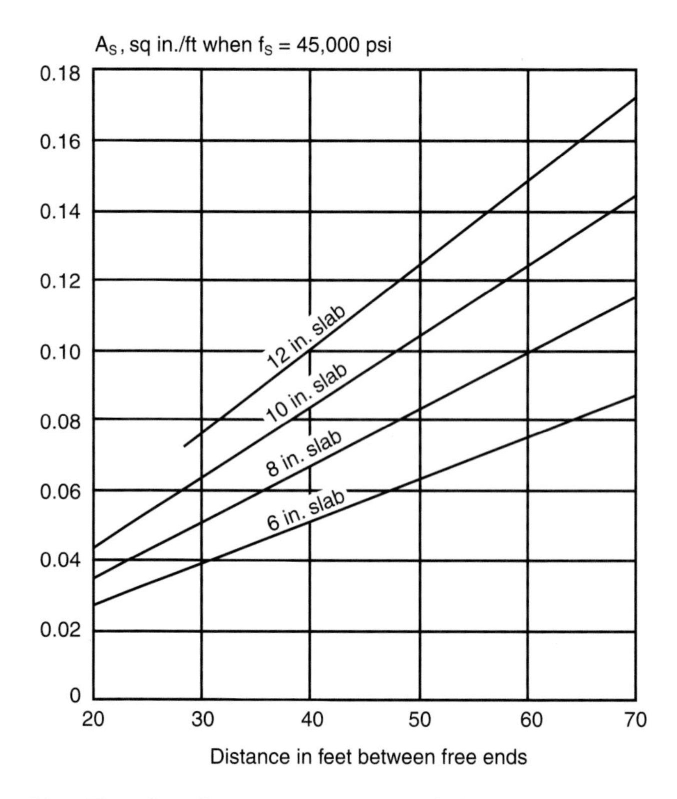

Note: The values shown in Fig. 6-14 were calculated by the subgrade-drag method. A working stress value of 0.75 times the yield strength was used because the consequences of a reinforcement failure are much less important than in normal reinforced concrete structural work.

Fig. 6-14. Selection chart for distributed steel.

(ACI 360). Based on experience and available information, a South African handbook on concrete floor design and construction recommends a minimum distributed steel percentage of 0.1% for high-yield steel bars and welded-wire fabric and 0.2% for deformed mild steel bars (Marias and Perrie 1993).

Alternative Methods for Determining Amount of Distributed Steel

Researchers at the University of Texas (Kunt and McCullogh 1990) have proposed a modified subgrade-drag formula for determining the required amount of distributed steel. Based on experimental results from studies of subbase friction, they developed the following formula, which better represents the actual frictional resistance of a range of available subbase types:

$$P_s = (600*L*T)/D*f_s \qquad (4)$$

where P_s = ratio of steel area to concrete area

L = spacing of transverse joints, ft

T = frictional resistance, psi

D = pavement thickness, in.

f_s = maximum allowable steel stress, psi

For a granular subbase, the frictional resistance is taken as 3.37. This formula yields significantly higher reinforcement ratios than does the original subgrade-drag equation.

Advantages of Higher Distributed Steel Quantities

Higher reinforcement ratios (than those calculated by the traditional subgrade-drag equation) provide an additional safety factor against crack widening. For instance, the steel requirement in short slabs may be doubled if a doweled joint freezes, thus doubling the spacing between working transverse joints. Highway researchers attribute the poor performance of some lightly reinforced jointed concrete pavements to steel yielding caused by malfunctioning doweled joints.

A typical amount of drying shrinkage might be about 200 to 300 millionths for a slab that contains an average amount of reinforcement. If the quantity of reinforcing steel were appreciably increased (so that the slab was considered "heavily reinforced"), it might provide enough restraint to reduce the shortening effects of drying shrinkage.

There are also construction advantages when using higher reinforcement ratios. Light gauge roll fabrics are often used for lightly reinforced slabs. There is no easy way to keep the fabric flat after it has been unrolled. Some of it bulges up near the top surface, and most gets stepped on and driven to the bottom of the slab, where it is of little use in holding cracks tightly closed. Reinforcing bars or sheets of welded-wire fabric with larger wires can be chaired to the correct height. And the spaces between bars or wires can be made larger to allow workers to step into openings, thus reducing the risk of steel displacement.

FLOORS WITHOUT CONTRACTION JOINTS

In some cases, contraction joints may be unacceptable to the building owner. Or the owner may prefer to eliminate contraction joints because of reduced floor and forklift maintenance costs. Three methods for building these floors are suggested:

- post-tensioned slabs
- continuously reinforced slabs
- shrinkage-compensating concrete slabs

Post-Tensioned Slabs

Post-tensioned prestressed slabs can be used to greatly reduce the number of floor joints in commercial and industrial floors. For building dimensions of about 60 m (200 ft) or less, the joints located near the periphery of the floor area can accommodate both slab end movements and post-tensioning anchorage. For larger facilities, post-tensioning anchorage and slab end movements have to be accommodated at interior floor locations (joints). Load-transfer dowels are installed at these joints, which are also detailed to handle wheeled traffic.

Post-tensioning a slab induces a compressive stress in the concrete that increases its modulus of rupture (flexural strength). This additional flexural strength allows a reduction of the floor thickness. However, reducing the floor thickness could result in greater deflections under load, depending on the subgrade support. In all cases, thickness design for post-tensioned slabs is based on the condition "interior slab loading" (see Chapter 5).

In most instances, post-tensioned slabs are installed over one or two layers of polyethylene sheeting to reduce subgrade friction. Tendons are located at about slab mid-depth with the ends placed slightly higher (one-third depth from the top) to counteract potential upward warping of the slab ends.

Tendons are anchored at slab ends. Post-tensioning is done in two or three stages as the concrete gains strength. The post-tensioning is initiated as early as possible to minimize random slab cracking. Each level of post-tensioning applied through anchor bearing plates should be balanced with the concrete strength to avert shear or other bearing failures.

To accommodate the post-tensioning equipment and activities, gap slabs are provided between ends of adjacent prestressed slabs or ends of slabs and building components. A gap slab is shown in Fig. 6-15. Joints are installed as part of the gap slab construction to accommodate slab end movements.

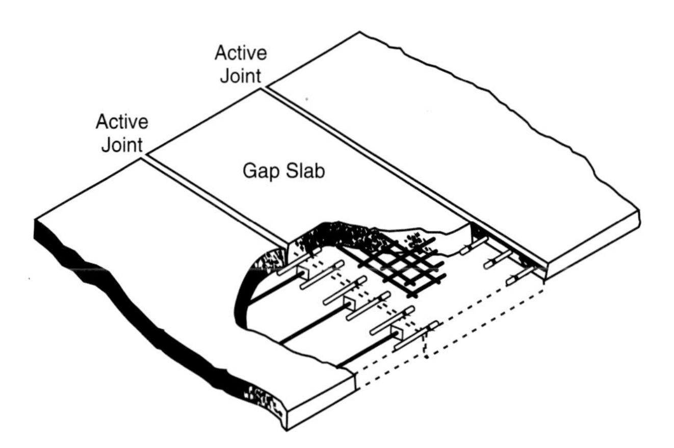

Fig. 6-15. Gap slab.

Joints at ends of long post-tensioned slabs should accommodate concrete horizontal dimensional changes due to long-term drying, creep, and ranges of facility temperature operating conditions. Joint widths should be selected to be compatible with lift-truck wheel types and diameters. The potential amount of joint widening for a post-tensioned slab is the sum of drying shrinkage (d_s), concrete creep (d_c), and temperature contraction (d_t). The potential joint widening can be reduced when an active joint is installed at both sides of a gap slab.

Most of the joint widening is expected to occur within one year after concrete placement. Beyond that time, changes in joint width for mature post-tensioned floors depend primarily on facility operating conditions, which could fluctuate as much as 11°C (20°F) as a result of wintertime heating.

Post-tensioned slabs must be designed by a structural engineer. Such a design involves choosing sizes and spacings of post-tensioning tendons or bars and strength of the steel wire. Tendons (strands) generally have diameters of 9 mm, 11 mm, and 13 mm (3/8 in., 7/16 in., and 1/2 in.) for slabs on grade. Spacings for tendons generally range from about 500 mm to 1000 mm (20 in. to 40 in.). Ultimate strength of the steel is 1860 MPa (270 ksi). The steel strands are typically tensioned to 80% of tendon ultimate capacity.

In addition to tendon anchorage installed at the ends of the post-tensioned slabs, load transfer dowels are installed between the ends of post-tensioned slabs and the gap slabs. Dowels are typically placed at 300 mm (12 in.) on centers at slab mid-depth.

Gap slabs typically contain reinforcement to hold potential cracks tightly closed. This reinforcement is placed at one-third depth from the top of the slab.

Some opening of joints is expected where the post-tensioned slab meets the gap slab. Tolerable joint widths at ends of post-tensioned slabs depend on types of lift-truck wheels that will pass over the joint. The objective should be to minimize wheel ingress into the joint space to prevent jarring and excessive vehicle maintenance.

A semi-rigid epoxy filler, as recommended for plain slab-on-grade joints, should be considered for filling active joints exposed to lift-truck traffic. The joints at the ends of new post-tensioned slabs will continually widen, often to a significant degree. To put a floor into service faster, the joint should be filled prior to introduction of traffic and should be repaired or replaced when widening makes the filler ineffective in spall protection. For mature installations, the movement will result only due to temperature variations and should require less repair. Fillers should be installed flush with the floor surface. Joints at ends of prestressed slabs that are not exposed to hard-wheeled traffic can be sealed with an elastomeric sealant.

Post-tensioned slabs may be used throughout a structure. Another option is to use them only in dedicated floor areas, such as for aisles between high storage racks having stringent flatness and levelness criteria. For relatively narrow post-tensioned slabs, the prestress may be needed only in the direction parallel to the slab's longitudinal axis. However, slab width should not exceed the contraction joint width criteria used for plain slabs. Post-Tensioning Institute publications should be consulted for further information on post-tensioned floors (PTI 1983, PTI 1990, PTI 1996).

Continuously Reinforced Slabs

Using relatively large amounts of continuous reinforcement is another way to control cracking in floors with no contraction joints. Continuous reinforcement is sometimes used in concrete pavements. At reinforcement percentages ranging from 0.5% to 0.7% of the slab cross-sectional area, the amount of steel is about 5 to 7 times greater than that used in conventionally reinforced slabs with "distributed" steel. Most floors that contain continuous reinforcement are designed with about 0.5% steel.

The high percentage of reinforcing steel does not prevent cracking. Instead, closely spaced, narrow cracks form. Crack spacings range from about 0.6 m to 1.2 m (2 ft to 4 ft). Because the cracks are narrow, load transfer occurs by aggregate interlock.

For long narrow slabs, such as aisles, the reinforcement is placed parallel to the longitudinal slab axis. The reinforcing bars are positioned at about one-third the slab depth below the slab surface and this should be 1.75 of the clear distance below the slab surface. Spacings between longitudinal bars should be about 150 mm to 200 mm (6 in. to 8 in.). As an example, for a 150 mm (6 in.) thick floor that contains 13-mm (No. 4) diameter deformed bars at 150-mm (6-in.) spacings, the amount of reinforcement is 0.55 percent.

Amounts of reinforcement in the slab transverse direction should be sufficient to avert longitudinal cracks. Transverse reinforcement can be dimensioned on the basis of the drag formula as presented on page 60.

Slab ends must be restrained for good floor performance. Restraints of slab ends to drying shrinkage and temperature contraction are essential to the formation of closely spaced cracks. In the absence of slab end restraints, crack spacing increases and cracks will widen near slab ends. (Observation of continuously reinforced highway pavements has shown that if slab ends are not restrained, roughly 46 m [150 ft] of the slab near the ends will exhibit movement and increased crack spacings.) Slab end restraint is typically provided in one of two ways. The slab ends are either tied into building foundations or into special below-grade bulkheads by means of reinforcement. Continuously reinforced concrete floors should only be tied to peripheral foundation walls that are designed to carry the horizontal thrust resulting from floor expansions. Even narrow cracks can fill with incompressibles over time. With rigid connections to walls and no room for movement, serious foundation wall displacements could result.

Thickness design for continuously reinforced slabs can be accomplished by using the design charts for the case of interior slab loadings found on pages 38 and 39. Where significant traffic or rack post loads are at or near longitudinal construction joints, the joints should be butt-type joints with load transfer dowels.

Shrinkage-Compensating Concrete

For large slabs, those ranging from about 900 m^2 to 1400 m^2 (10,000 ft^2 to 15,000 ft^2), constructed with a shrinkage-compensating cement concrete, no intermediate control joints are needed. Construction joints are installed along the outside edge of each slab. The shrinkage-compensating cement concrete works together with the slab reinforcement to avert random intermediate slab cracking.

The following description of shrinkage-compensating concrete shows that the concept is simple. After concrete is placed, and initial hardening occurs, a bond develops between the concrete and the reinforcing steel, and shrinkage-compensating cement undergoes a slight expansion. During moist curing, as the concrete expands, the steel restrains the expansion, and the steel is consequently put into tension. Thus, at the end of moist curing a quasi-prestress exists within the slab. When the concrete subsequently dries and contracts, the tensile forces in the steel are lost. The benefits of concrete expansion restraint by the reinforcement during the moist curing period are twofold:

- Total drying shrinkage is reduced.
- The concrete gains tensile strength before contraction restraint stresses develop.

Thickness design for the large shrinkage-compensating cement concrete slabs is based on interior loading conditions. Design charts (nomographs) presented in this publication can be used to determine thickness.

The amount of reinforcement needed to restrain concrete expansion is 0.15% of slab cross-sectional area of the concrete. The reinforcement should be installed at about one-third of slab depth below surface and terminated at construction joints. Structural slabs, of course, would still contain the required two mats of steel reinforcement.

Slab end movements due to long-term drying shrinkage, if any, depend on effectiveness of the shrinkage- compensating system. Slab end movements due to exposure of the floor to temperature variations are accommodated at the construction joints. Considerations are similar to those discussed for post-tensioned slabs for active joints.

Though the concept of shrinkage compensation is well developed, the design and construction of such floors require special materials, knowledge, and experience with this type of system.

Stoppage of Work

Whenever continuous concrete placement will be interrupted for 30 minutes or more, a bonded construction joint should be inserted to avoid the formation of a cold joint. A bonded construction joint in a plain slab is a butt-type construction joint with tiebars. Tiebar sizes and spacings are shown in Tables 6-2a and 6-2b. Any reinforcement in the slab is continuous through the bonded construction joint. The bonded construction joint should not be used if it is located at a contraction joint alignment.

Construction joints, except for "emergency" bonded construction joints, will accommodate slab end movements, and thus function similar to contraction joints. All slab reinforcement should be terminated at the joint lines. Load transfer dowels should be smooth bars and should be installed parallel both to the floor surface and slab centerline.

CHAPTER 7

CONCRETE PLACEMENT AND FINISHING

Concrete should be placed and finished by skilled workers who have experience with concrete floor construction. Finishing, especially, requires skilled cement masons for best results. Producing a finished surface usually requires three stages:

1. concrete placement, compaction, and truing (to a rather rough surface) of the struck-off or screeded surface by the use of a hand-screed or vibratory screed
2. shaping and smoothing the surface by bullfloating or darbying, followed by straightedging
3. final compaction and smoothing by steel-troweling with a hand or power-driven trowel

There are three basic finishes for a concrete slab surface: screeded, floated, and troweled.

A *screeded* finish involves the least amount of work. Immediately following consolidation, surplus concrete is removed by striking off the surface using a straightedge. Certain types of construction require no more than a screeded finish. The name of the finish is derived from the guides, called *screeds*, that determine the elevation of the surface. Edge forms can be used as screeds. Placing a narrow row of concrete at a certain elevation can serve as a wet screed, though this method does not afford tight control over finished elevations.

A *floated* finish is normal for outdoor slab surfaces. After screeding, the concrete begins to stiffen and the water sheen disappears from the surface. At this point, floating begins. Floating should work the concrete no more than necessary to produce a surface that is level, uniform in texture, and free of foot and screed marks. If a troweled or broomed finish is to be applied, floating should leave a small amount of mortar on the surface. There should be no excess water. The mortar permits effective troweling or brooming. For interior slabs exposed to freezing and thawing during construction, and for all outdoor slabs, the finishing procedure should not remove entrained air from the mortar surface.

A *troweled* finish is used on interior floor slabs. Steel troweling improves the cosmetic appearance and provides a surface that is stronger, more wear resistant, and easier to clean. The steel troweling may be omitted if a coarse texture is desired. Two conditions must be met to properly time the troweling: the moisture film and sheen should have disappeared from the floated surface, and the concrete should have hardened enough to prevent an excess of fine material and water from being worked to the surface. Steel troweling is performed with a firm pressure that will transform the open, sandy surface left by floating into a hard, dense, uniform surface free of blemishes, ripples, and trowel marks.

Different degrees of troweling yield different finishes. A very fine-grained swirl finish is the result of light troweling. To create this nonslip finish, the trowel is held flat and moved over the surface in a circular motion immediately after the first regular troweling. A very smooth hard finish is the result of repeated troweling. If the hard steel troweling is repeated until the surface has a somewhat polished (glossy) look, a burnished surface will result. This special finish provides added resistance to abrasion and wear. See Chapter 8 for detailed information about producing burnished floor finishes.

Industrial and commercial (interior) floor slabs should have a troweled finish, but there are no compelling reasons to trowel finish an outdoor slab. Texture is better for outdoor surfaces, and steel-trowel finishes provide little traction when wet. If a light broom finish is wanted, the surface should be floated, troweled, and then broomed. If a rough broom finish is desired, the surface should be floated, followed by rough brooming. (In this case, the troweling step is skipped.) Another reason to avoid troweling outdoor slabs is that densification by troweling may eliminate entrained air from the concrete surface, leaving it vulnerable to scaling.

PREPARATIONS FOR PLACING CONCRETE

Subgrades and Slab Thickness Tolerances

Good floors begin with well-prepared subgrades. Many floor problems can be traced to poor subgrade preparation.

The subgrade should be uniform, firm, and free from all sod, grass, humus, and other rich organic matter, as these materials do not compact. See Chapter 2 for a discussion of soil properties as they relate to the proper type of subgrade and preparation for concrete floors on ground.

For floors with heavy loads—post loads or direct storage on floors—special considerations regarding soil consolidation, densification, and/or subsidence are mandatory during design and construction.

Concrete should be placed on a level, uniform surface. Placement will either be directly on the subgrade or on the subbase. The *subgrade* is the natural in-place soil; the *subbase* is generally a compactible fill material that brings the surface to the proper grade. The subgrade/subbase must be brought to within required tolerances at the specified grade. A reasonably accurate, level subgrade will ensure that the correct thickness of concrete (and subbase if needed) is placed. If the subgrade or subbase surface is uneven, concrete will be wasted, and the potential for random cracking will be increased.

Using laser alignment tools permits measurements of the graded surface to be taken quickly. A scratch template can be used as a check to reveal high and low spots. For even faster grading with a high level of accuracy, laser-guided grading boxes can be used (see Fig. 7-1). This equipment is capable of fine grading the base to a tolerance of ± 6 mm (1/4 in.), which can save substantial amounts of concrete on large slab areas and also improves the surface flatness and levelness of the floor (Basham 1998). After the subgrade/subbase has been leveled, it should be compacted. The subgrade/subbase should be moist but not saturated when concrete placement begins (Fig. 7-2).

Section 4.4.1 of ACI 117-90, *Standard Specifications for Tolerances for Concrete Construction and Materials,* gives acceptable thickness tolerances for cast-in-place concrete slabs. ASTM C 174, *Standard Test Method for Measuring Thickness of Elements Using Drilled Concrete Cores.*

For slabs with cross-sections up to 300 mm (12 in.) thick, the limits are:

+9.5 mm (+3/8 in.) and -6 mm (-1/4 in.)

For slabs thicker than 300 mm (12 in.), but not more than 900 mm (3 ft) thick, the limits are:

+13 mm (+1/2 in.) and -9.5 mm (-3/8 in.)

The thickness tolerance statement shown above applies to structural slabs—whether they are on grade or suspended. From floor load-carrying capability considerations, exceeding the design thickness is not detrimental. Constructing slabs thinner than shown on the plans, however, can have a significant impact. The stated -6 mm (-1/4 in.) tolerance can raise flexural stresses by about 10% for 125-mm (5-in.) thick slabs. For thicker slabs, the percent increase in flexural stress is not as great. In theory, it would be possible to reduce the tolerance to 3 mm (1/8 in.) for thinner floors (125 mm to 150 mm [5 in. to 6 in.]) with significant loads, but it would be difficult to construct a subgrade to such a tight tolerance. Within reason, increased floor thickness from dimensions shown in contract documents is acceptable. One caution is that sudden thickness increases from ruts or holes in the subgrade/subbase should be avoided, as sudden differences in (concrete) thickness can lead to localized stresses and cracking. Where floor elements are thickened, the transition slope should be shallow (no more than 1 vertical: 10 horizontal).

Fig. 7-1. A laser-guided grading box can quickly and accurately level the subgrade. (69660)

Fig. 7-2. Subgrade and/or subbase should be moist when concrete is placed. (69618)

Expansive Soils

Expansive soils under slabs often lead to problems. Excessive swelling can cause subgrade heaving; differential shrinkage can cause subgrade settlement. Without uniform support, a floor slab may distort and crack. The harmful effects of expansive clay soils can sometimes be made less severe if the upper 200 mm to 300 mm (8 in. to 12 in.) is stabilized by mixing in cement or lime followed by compaction. For soil conditions like those found in parts of the southeastern United States, special cardboard forms can be used to create blockouts beneath concrete floors. These voids serve as outlets for soil expansion and prevent unwanted pressure on the floor.

Subbase or Cushion

If a floor on ground will not be subjected to moving loads, it can be built without a subbase. However, a subbase is frequently used as a leveling course for fine grading and as a cushion that will equalize minor surface irregularities and contribute to uniform support. It also may serve as a capillary break between the floor slab and subgrade, depending on its thickness.

For jointed slab-on-grade floors exposed to moving loads, a subbase sandwiched between the concrete slab and the subgrade is mandatory for all but well-graded sand and gravel subgrades (AASHTO classification A-2-4 or better). The minimum subbase thickness is 25 mm (1 in.). If the subbase is also acting as a capillary break, its thickness is commonly 100 mm (4 in.) or more.

A subbase composed of open-graded crushed material or gravel should receive a choker course of fine aggregate to fill surface voids. This filler course does a number of things. The choker course helps to bind, or tighten, the surface of the fill during compaction. The fine aggregate particles rest in the voids between larger aggregate particles, so the surface is relatively smooth. This reduces friction between the slab and subbase, minimizing restraint that can contribute to random cracking. The smooth surface is also less likely to rip a vapor barrier placed over it.

The subbase is fine graded, or trimmed, and compacted to maximum density before placing concrete. Granular subgrades and subbases are compacted with a vibratory roller, a flat-plate vibrator, or a tandem roller. Hand tampers can be used in confined places. Sheepsfoot rollers or hand-held rammers are the best method of compacting cohesive subgrades. On large floor projects, compaction equipment similar to that used in highway or airport construction can be used.

Vapor Retarder

Many of the moisture problems associated with floors on ground can be minimized or eliminated by proper preliminary grading, correct selection of fill or subbase materials, and installation of a vapor retarder. The flow chart in Chapter 3 along with the following discussion will help guide designers and specifiers in the selection and placement of vapor retarders.

A vaporproof membrane should be placed beneath all concrete floors on ground that receive an impermeable floor finish. These "nonbreathing" finishes include resilient tile, rubberized carpet backing, epoxy mortar surfacings, and urethane coatings. Some breathable coverings, such as wood flooring and cork tile are also affected by moisture.

Vaporproof membranes are required for floors in humidity-controlled areas, interior locations where the passage of water vapor through the floor is undesirable. This includes warehouses having storage areas for cardboard boxes, pallets, and other objects. It also includes living or work areas where furniture sits directly on a floor. Without a vapor retarder in place, moisture can condense beneath these items, causing moisture damage: dampness, mildew, and rot.

Good quality, well-consolidated concrete at least 100 mm (4 in.) thick is impermeable to the passage of *liquid water* from the ground, unless the water is under considerable pressure; but concrete several times that thick is not impermeable to the slow passage of *water vapor*.

Water vapor normally passes through the concrete and evaporates at the slab surface. In the absence of vapor retarders below floors, excess moisture can pass through the concrete, resulting in adhesive failures and delamination of vinyl floor coverings (see Figs. 7-3 and 7-4). Urethane coatings and epoxy surfacings may become unbonded due to water vapor pressure, but this generally only occurs if the coating becomes emulsified again or if it did not properly bond when first applied. If good bond develops at the time of application, water vapor pressure should not be great enough to lift the coating from the surface.

Vapor retarder materials with a permeance of less than 0.20 metric perms (0.30 perms) are suitable for floor-on-ground construction. Acceptable materials are available in preformed sheets or mastics that will resist deterioration and punctures from subsequent construction operations.

Vapor retarder material placed directly under the slab also functions as a slipsheet to reduce subgrade drag fric-

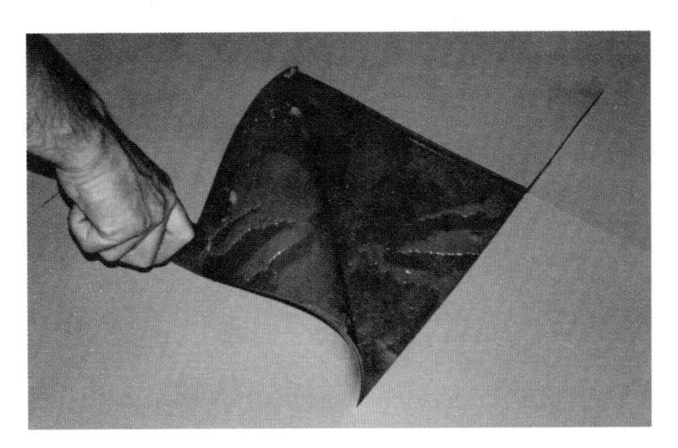

Fig. 7-3. Adhesive failure on a concrete slab. (68129)

Fig. 7-4. Delamination of a floor covering installed over a concrete floor. (68130)

tion, permitting freer slab movement and reducing cracking in the slab. However, vapor retarders may contribute to the slab's upward warping (curling) and aggravate problems of plastic cracking. When a vapor retarder is used, a 75-mm (3-in.) thick layer of granular, self-draining compactible fill can be placed over the vapor retarder to minimize the above problems. The layer also protects the vapor retarder from puncture during construction operations. Care should be exercised that sharp stones from above or below the vapor retarder do not pierce the membrane.

Insulation Under Slabs

For floors containing heat piping or other heating provisions, insulation below the floor can provide energy savings. Insulation below concrete slabs is mandatory for cold storage, ice-skating rinks, and freezer floors. The insulation averts frost penetration into subbase and subgrade soils and prevents frost heaving of these materials. For freezer floors and ice rinks, the insulation is located between the concrete slab bottom and a mudslab. Mudslabs are generally constructed by placing a lean concrete mix. Heat return ducts are encased in the mudslab to prevent subgrade freezing and heaving that could occur in spite of the insulation below the floor. Mudslabs also provide a stable working surface.

Under-slab insulation has to perform two functions: it has to insulate the slab and it has to provide support to the slab both during and after construction. Insulating properties are based on the material itself and the thickness used. The support provided is known to vary with insulation board density and is given as the bearing value (equivalent modulus of subgrade reaction) determined by a plate load test (ASTM Designation D 1196). Insulation below floors should not be allowed to absorb water, as the insulating quality is lost with increases in moisture content.

In cold climate regions, insulation below floors on ground is used at floor edges adjacent to exterior foundations or walls. The insulation is placed inward for a distance of about 600 mm to 900 mm (24 in. to 36 in.). The insulation may also be used vertically along the foundation wall to aid in heat retention.

Edge Forms

Edge forms or sideforms used along longitudinal casting lane sides are set to the proper floor surface elevation and checked. Grades should be marked on stakes at floor interiors and along foundation walls at slab peripheries to show top-of-finished-floor surface elevations. When the sideforms are set, the slab thickness should be checked. Between forms, the thickness measurement should be made from a straightedge laid across the forms down to the top of the subgrade. If wet screed methods are used for surface control, string lines should be used from marked stakes and walls to measure downward to the subbase/subgrade surface. High spots should be removed and low spots filled with the same material. The surface should be choked and recompacted as needed prior to placing concrete.

The surface-finish tolerance of the floor slab depends on careful setting of the edge form elevations. Temporary intermediate screed guides are placed between the edge forms, and their elevation is established using a leveling device set on the edge form. This provides close control of surface elevation.

Concrete placement for floors is typically done in lanes; the sequence may be either alternate lanes or adjacent lanes (see Fig. 7-5). If the method chosen is alternate lanes, sideforms are set on both sides of the lane. One lane is placed, the adjacent lane is skipped, and the next lane is placed. Once two lanes have been placed, the intermediate lane may be placed between them. With good elevation control on the two outside lane placements, the lane between them will be at the proper elevation. If adjacent, rather than alternate, lane placement is chosen, then after

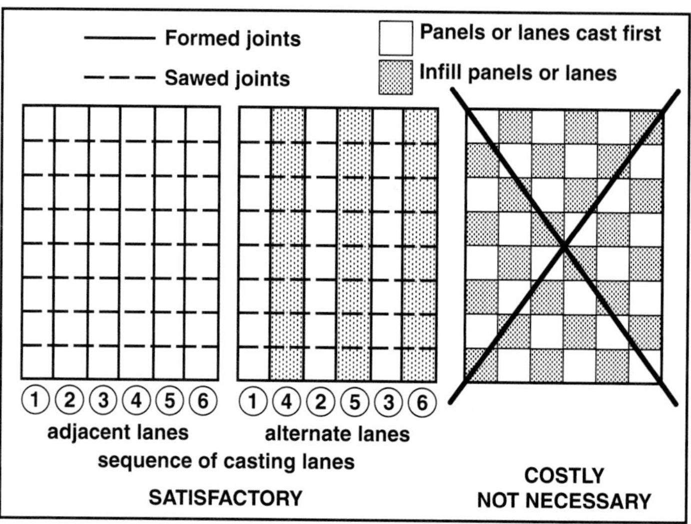

Fig. 7-5. Sequence for casting concrete floors on ground.

the first lane has been placed, only one sideform is needed for subsequent lane placements.

Checkerboard methods for concrete placement should not be used. Though sometimes used in the past in an attempt to minimize cracking from shrinkage, this method is labor intensive and does not provide any benefit because construction schedules are too fast to allow shrinkage to take place prior to the placement of the alternate slabs.

Edge forms and screed guides must be supported firmly by wood or steel stakes driven solidly into the ground to prevent any movement during mechanical strikeoff and consolidation. Loose edge forms cause uneven floors. Temporary screed guides that are not positively positioned can be displaced, causing uneven floors. All forms should be straight, free from warping, and of sufficient strength to resist concrete pressure without bulging. Edge forms should be full slab depth and continuously supported on the compacted subgrade or subbase (see Fig. 7-6); for this reason steel road-type forms are preferred. Vibratory strikeoff and compaction equipment slides easily on a steel surface. A release agent eases form removal.

When a continuous vapor retarder or barrier is required under the floor slab, special care is needed to place the sheet material correctly under the edge forms before the pins are driven.

In order to minimize construction costs when placing very flat or superflat floors, flatness characteristics should be specified only as strict as required for the operations that will take place on the completed floor. In facilities where narrow aisles are used for high-rack storage, the truck-dedicated lanes (aisles) should be placed first, with close attention paid to surface elevations. These lanes can then serve as screed support for strikeoff of the infill lanes dedicated to support racks, which require a lesser surface profile quality. In all cases, a properly graded subgrade meeting tight tolerances makes it easier to obtain a more flat and level floor surface.

Fig. 7-6. Edge forms and screed guides must be solidly supported to prevent any movement during strikeoff and consolidation. (69619)

Positioning Reinforcing Steel and Embedments

Reinforcing bars or welded-wire fabric (wire mesh) are used in floor slabs for crack control. In order for reinforcing bars or mesh to bond properly with the concrete, they should be free of mud, oil, or other coatings that would adversely affect the bonding capacity. The reinforcement should be placed in the upper portion of the slab—50 mm (2 in.) below the surface is suggested—and if it is to function properly for crack control, it must be held securely in that position during concrete placement. Also, it is imperative that the steel be discontinued at all contraction joints.

Embedments describe items other than reinforcement that are located partially or completely within the concrete slab. This includes fully embedded circulation coils, such as coolant ducts for ice rinks and heating ducts for heated floors; drains for the floor surface; and pipes that pass completely through the slab. Embedments can be supported on chairs, slab bolsters, or small concrete bricks and blocks. Sand plates should be used with reinforcement or ducts supported above sandy subbase material and above vapor retarders and insulation. Supports should be strong enough and the spacing close enough to endure imposed loads from the concrete placing crew without displacement of the embedded item.

Support accessories for wire mesh must be more closely spaced than for reinforcing bars (see Fig. 7-7). The reinforcement and supports should be able to support the concrete placement crew without permanent downward displacement. The practice of laying the mesh on the subgrade/subbase before concrete is placed, then "hooking it up" into position during concrete placement is not recommended. Alternately, some people claim that wire fabric can be placed in the slab by "walking it in." After the con-

Fig. 7-7. A rebar mat is supported to hold it in place while concrete is being placed. (69686)

crete is placed full depth and struck off, mesh is placed on the surface and the finishing crew carefully walks on it, forcing it into the concrete. This method requires a low-slump concrete, and even then accurate positioning of the steel fabric is difficult. Walking the steel in is not a recommended method for placing mesh. Neither method—hooked up or walked in—gives assurance that the mesh will be located at the proper depth in the concrete nor that it will remain flat.

Flat-sheet mesh is preferable to rolled mesh. It can be placed by the sandwich method in two-course work. This involves first placing a layer of concrete struck off 50 mm (2 in.) below the finished grade. The mesh is laid on this layer, and then the top layer of concrete is placed. Work must be completed quickly so that the top layer is placed while the bottom is still plastic. Regardless of slab thickness, vibration with spud vibrators (internal slab vibration) is recommended when the sandwich placement method is used. Two-course work requires more labor than single-course methods.

Wherever dowels or tiebars are placed in joints, they should be installed in sideforms or in dowel baskets at contraction joint alignments. Contraction joints should be marked at tops of forms to assure proper joint location. The usual methods of creating a contraction joint are to place joint formers during concrete placement or make sawcuts after concrete finishing. One-half of each dowel length should be oiled to prevent bond to the concrete. Dowel alignment with the horizontal and vertical planes should be checked and corrected, as needed, immediately prior to concrete placement.

Tolerance for Embedded Items

Random cracks can occur when concrete is restrained against contraction or can occur as a result of slab warping. Temperature and shrinkage reinforcement (used to limit the width of random cracks) should be placed close to the top surface, yet deep enough to avert surface cracks over the reinforcement. Place welded wire reinforcement 50 mm (2 in.) below the surface or in the upper one-third of the slab thickness, whichever is closer to the surface. Reinforcement should extend to within 50 mm (2 in.) of the slab edge (joint). In chloride environments, concrete cover for corrosion protection governs steel placement.

Round or square smooth-surface load-transfer dowels should be 360 mm to 460 mm (14 in. to 18 in.) in length, with approximately one-half of the dowel extending into either adjoining slab. Dowels should be located parallel to the slab surface; for sloping slabs, the dowel should be installed at the same slope as the slab surface with a vertical accuracy of ± 3 mm (1/8 in.).

Embedment of heating ducts or coolant conduits in floors should be at a clear depth of 50 mm to 75 mm (2 in. to

Fig. 7-8. (top left) Concrete is most easily deposited into forms right off the truck by means of the chute. (bottom left) Buggies or wheelbarrows allow moving small loads of concrete to hard-to-access areas on grade. (top right) A crane with a bucket attachment can place concrete from a central location and can be used for at-grade work. (bottom right) A concrete pump is a very common method of placing concrete, and the discharge is easily moved around the site. (69661, 54088, 69687, 69424)

3 in.) below the surface to provide uniformity of heating or freezing at the concrete surface. Electric resistance heating cables are embedded 25 mm to 75 mm (1 in. to 3 in.) below the surface. At least 25 mm (1 in.) of concrete below the piping or cables is needed, but twice that amount is preferable.

CONCRETING PROCEDURES

Tools for placing, consolidating, and finishing concrete floors are largely mechanized, though some amount of hand labor is generally necessary for striking off, compacting, and finishing. Large jobs require more equipment, and small jobs may have restrictions based on access to the work area.

Floors cast under the protective cover of a tight roof with the walls of the building in place are apt to be better constructed than floors cast without overhead protection. Outside work can be adversely affected by the weather (sun, heat, cold, wind, and rain) and have a greater risk of problems such as cracking, crazing, and curling. The main precaution for indoor construction is to vent heater exhaust to the outside whenever heaters are used to protect the concrete from freezing. Otherwise, poor air quality could lead to health concerns for workers and surface dusting of the floor.

Placing and Spreading

Rutted or marred subgrade or subbase surfaces must be regraded and recompacted before placing concrete. There are several ways to place fresh concrete where it is needed, including directly from a truck mixer's chute, by buggy on wooden ramps, by crane with bucket, by conveyor belt, or by pump (see Fig. 7-8 and PCA 1988). Concrete placement should start at corners and proceed away from the corners. Continuous, uninterrupted placement is desirable. Concrete should be placed as closely as possible to its final position and slightly above the top of the edge forms or screed guides. It is then spread with shovels, special concrete rakes, or come-alongs. Vibration should not be used to move the concrete laterally. Air trapped in the concrete during mixing and placing must be removed by consolidation.

Striking Off and Consolidating

Concrete is brought to its initial level and surface by the operations of strikeoff, consolidation, and darbying or bullfloating. Striking off, or screeding, is the action of leveling the top surface of concrete when it is first deposited into the forms. Where surface tolerance is important, the strikeoff lane width should be limited to 6 m to 7 m (about 20 ft to 24 ft), even less for superflat work. The limited width makes it possible to use a crossrod or straightedge more accurately.

Strikeoff can be done by a variety of equipment, varying from simple methods (straightedges and roller screeds), to vibratory screeds, to high-tech laser screeds (see Fig. 7-9). Straightedges and vibratory screeds ride on top of the side forms, but laser screeds need no screed rails. Laser screeds are capable of leveling the surface of fresh

Fig. 7-9. (top) A template, or straightedge, can be used to strike off and level the concrete. (second and third photo) Strikeoff can be done with a vibratory screed. (bottom) A laser screed quickly places concrete to tight tolerances. (69662, 69663, 69620, 69664)

concrete to tight tolerances because the elevation is constantly and automatically monitored and adjusted. These machines are mobile and capable of moving quickly from one location to another to speed the strike off operations.

Surface vibration is considered sufficient consolidation effort for slabs of up to 200-mm (8-in.) thickness; when slabs are thicker than this, internal vibration is required. Internal vibration is also needed for floors that contain heavy reinforcement, load transfer dowels, or conduit. When a thick slab has been compacted by internal vibrators, final compaction of the surface should be done with a beam- or truss-type surface vibrator.

Strikeoff and consolidation must be completed before any excess water bleeds to the surface. The secret of proper strikeoff and compaction is to maintain an adequate surcharge of concrete at the screed face; a 150-mm (6-in.) thick slab needs a surcharge of about 25 mm (1 in.). If power-driven equipment is used properly to strike off low-slump concrete, the surface will be ready for edging and finishing with power floats and trowels with very little handwork necessary.

The initial strikeoff and consolidation of the concrete will have a greater effect on surface tolerances and levels than subsequent operations of floating and finishing.

Leveling

After strike off, the concrete surface is further leveled and smoothed to prepare the surface for subsequent finishing operations. Leveling should immediately follow screeding and be completed before any bleedwater is present on the surface (see Fig. 7-10). *Any finishing operation performed while there is excess moisture or bleedwater on the surface can cause surface defects.*

This leveling/smoothing is done with a bullfloat, a powerfloat, or a darby (see Figs. 7-11, 7-12). The purpose of the bullfloat or darby is to eliminate ridges and fill in surface voids left by striking off and consolidating. This is also referred to as cutting high spots and filling low spots. In addition, the bullfloat or darby slightly embeds the coarse aggregate.

Surface flatness and levelness requirements dictate the length of bullfloat that should be used; longer bull-floats yield flatter floors. For overall flatness of 20 or less, a 1.2-m to 1.5-m (4-ft to 5-ft) bullfloat can be used. For flatness values between 20 and 25, a 2.4-m to 3.0-m (8-ft to 10-ft) long bullfloat should be employed (see Table 7-2). For finishes with a flatness number greater than 25, a 3-m (10-ft) long highway straightedge must be used for the second straightening pass and for any subsequent passes.

Bullfloats and darbies are used for the same purpose but provide different advantages and yield slightly different results. Because bullfloats have long handles, they make it easier to reach large areas but harder to apply leverage. This makes achieving close tolerances more difficult and may adversely affect the flatness of a floor slab. Furthermore, the bullfloat may slightly depress or raise the

Fig. 7-10. A bleeding slab surface. (69665)

Fig. 7-11. Floating corrects small irregularities and smoothes out ridges. (69666)

Fig. 7-12. Darbying brings the surface to the specified level and is done in tight places where the bullfloat can not reach. (42241)

Fig. 7-13. A highway straightedge helps maintain tight tolerances during the early finishing operations. (69719)

Fig. 7-14a. Edging involves cutting concrete away from the forms with a margin trowel. (34064)

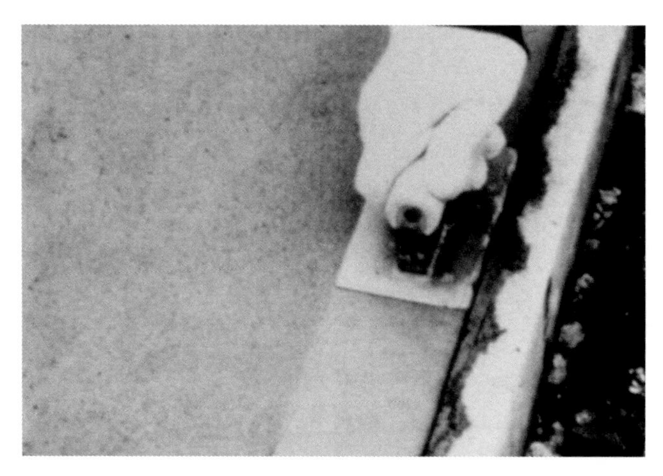

Fig. 7-14b. The second step in edging is running the edger to smooth and densify the corner. (68973)

surface near an edge form, which would then require correction—releveling the surface with a darby. As for darbies, their short handles make it easier to exercise control over the finishing operation, but make it difficult or impossible to reach far from the edges of the concrete slab.

Where close tolerances are required, a straightedge can be used to advantage. Minor surface irregularities and excess laitance are removed by scraping with a 2-m to 4-m (6-ft to 12-ft) straightedge (Fig. 7-13). It should be used with a smooth, continuous action to float the surface; a jerky, cutting action is used only for removing high spots. Each pass should overlap one-half the length of the previous pass. Surface smoothness should be checked as late in the finishing operations as possible but while the concrete is still plastic enough to permit surface corrections. To test the flatness, a checking straightedge is laid across the surface in successive positions.

On a typical construction project, a bullfloat or darby is used immediately following strikeoff. It improves the resulting floor finish. However, not all floor surfaces have strict requirements for flatness and levelness, and bullfloating or darbying after strikeoff is optional.

Edging

Edging may be required along isolation and construction joints. The purpose is to densify the concrete at the edge of the slab, making it more durable and less vulnerable to spalling and chipping. Edging should be completed before bleedwater comes to the surface. Before running the edger, the concrete should be cut away from the forms with a pointed mason's trowel or margin trowel (see Figs. 7-14a, 7-14b). An alternative to using an edging tool at construction joints is to lightly grind the edges with a silicon carbide stone after the forms are stripped and before the adjacent slab is placed.

At construction joints exposed to truck traffic with solid tires or hard casters, joint edges need lateral support to prevent spalling and chipping. No edging or stoning is done. Instead, a 25-mm (1-in.) deep sawcut is made at the construction joint to create a reservoir that will be filled with a semi-rigid filler.

Following edging, a slight stiffening of the concrete is necessary prior to floating. This stiffening requires time and is dependent on many variables including the concrete mix and the weather (temperature, humidity, and wind).

Floating

The concrete should be allowed time to stiffen before it is floated. Floating, whether it is done by hand or by machine, has four purposes:

- to depress large aggregates slightly beneath the surface
- to remove minor imperfections and even out humps and voids for a smooth surface
- to bring mortar to the surface in preparation for later finishing operations
- to keep the surface open so excess bleedwater and bleed air can escape

In general, three conditions indicate that concrete is ready for floating:

- the water sheen has disappeared from the surface
- the concrete will support the weight of a person with only a slight surface indentation
- mortar is not thrown by the rotating blades

The surface indentation should be no more than 6 mm (1/4 in.) (Fig. 7-15). The conditions will be met after a waiting period, which is sometimes difficult to determine. The length of time required for reaching the proper conditions varies; in cold weather, it might be 3 hours or longer, but in hot weather, the concrete may stiffen rapidly. Experience plays a big part in judging when the slab is ready for further finishing operations.

Floating concrete is now done almost exclusively by machine. Either a power float with rotating steel disc or a troweling machine equipped with float blades is used. *Caution: Pans must not be used too early.* Use conventional float blades or float shoes on the first pass. Using pans too soon could create blistering, especially where dry-shakes are used. Float blades should sit almost flat on the surface. Floating should start along walls and around columns and then move systematically across the surface, leaving a matte finish. The float should also be run along and over edge forms.

Marks left by the edger should be removed by floating. During the interval between power floating and the first power troweling, a steel hand trowel should be used along the edges to improve the surface and keep the concrete level with the side forms. As with floating, troweling should be done along and over edge forms.

Fig. 7-15. A shallow footprint is one of the indications that concrete is ready for finishing. (69429)

Finishing Flat and Level Floors

Finishing flat and level floors requires more attention to detail than normal finishing, but follows the same general steps. For successful installation of flat and level floors, forms should be set to the required surface elevation accuracy listed in Tables 7-2 and 7-3 (as presented under the discussion of floor surface tolerance). For levelness of 50 and better, the tops of forms should be checked with a straightedge, and high points should be planed. As levelness requirements increase, the distance between forms (lane width) should be reduced.

The flatness and levelness characteristics are better obtained through the use of vibratory screed concrete strikeoff and consolidation. Floors thicker than 200 mm (8 in.) or floors with embedments (such as load transfer dowels) require that spud vibrator consolidation be followed by vibratory screed strikeoff. The strikeoff surface should be checked with a straightedge, and strikeoff should be repeated as needed.

Highway straightedges are used for cutting floor surface high spots and filling low spots. Their use should follow initial smoothing with a bullfloat, waiting (for cessation of bleeding), and initial machine floating. As flatness and levelness criteria are raised (as listed in Tables 7-2 and 7-3) and placement lanes are narrowed, straightedging is done at 45° to the lane axis. If low spots are encountered, they are filled with additional concrete mix with the largest size coarse aggregate removed.

For superflat floors with F_F/F_L-values at 50 or higher, vibratory screed strikeoff is followed by several (hand) passes of a straightedge. Initial smoothing is accomplished with a wide bullfloat used at 45° to the lane axis. Following initial power floating, cutting and filling of high and low spots is done with a highway straightedge. Power troweling follows this. Cutting of high spots follows initial power troweling; there should be no low spots. After an additional waiting period to allow more concrete stiffening, more troweling is done to densify the surface for increased wear resistance.

Troweling

The purpose of troweling is to improve localized surface smoothness and to provide a hard abrasion-resistant surface for the floor. The troweling action densifies surface mortar. Both free water and air are expelled from the slab surface. For mortar near the surface, troweling reduces the water-cement ratio and increases its strength. When cores from troweled concrete slabs are examined, there is a color difference between the top 3-mm (1/8-in.) layer of surface mortar and the core body. The darker mortar layer is indicative of a lower water-cement ratio.

A power trowel is like a power float except that it is fitted with smaller, individual steel-trowel blades that can be tilted slightly to exert pressure on the surface. Generally, greater tilt will produce a smoother and denser surface.

Power troweling should start when the excess moisture brought to the surface by initial power floating has evaporated and the concrete is not sticky. The waiting time between floating and troweling depends upon the absence or presence of an admixture in the concrete and the atmospheric condition at the surface.

Power troweling should be done in a systematic pattern. Two or more passes frequently are required to increase the compaction of fines at the surface and give greater resistance to wear. Time must be allowed between each troweling for the concrete to stiffen and the water sheen to disappear. The tilt of the trowel blades should be increased with each pass to exert additional pressure as the concrete hardens. Each successive troweling should be made in a direction at right angles to the previous pass.

Jointing

Proper jointing can eliminate unsightly random cracks. Aspects of jointing that lead to a good job are choosing the correct type of joint for each location, establishing a good joint pattern and layout, and installing the joint at the correct time. Timing of the jointing operations depends on the tool used to make the joint. (Formed joints have to be made before troweling.)

Contraction joints can be made with a hand groover, a preformed insert, or a power saw; however, sawed joints are the only acceptable type for joints in floors subject to forklift traffic.

The joint groove, plastic insert, or sawcut should extend into the slab one-fourth of the slab thickness. Sometimes, a joint depth of one-third the slab thickness is specified. Some research indicates that a cut shallower than one-forth may be adequate if made at the proper time. Whatever the depth of groove, insert, or sawcut specified, it must be adequate to raise the tensile stress in the reduced slab section below the joint and thus induce a crack to form beneath the joint, where it will be inconspicuous.

There are few instances in industrial and commercial floor work when a hand groover is used to make joints; however, the groover's bit must be thin and deep enough to cut the slab one-fourth of the depth. If joints are created with hand tools, the control joints should be cut into the surface while the edging is being done or immediately after edging.

On large floors it is more convenient to cut joints with a power saw fitted with an abrasive or diamond blade (Fig. 7-16). Sawing should begin as soon as the surface is firm enough so that it will not be torn or damaged by the blade, usually within 4 to 12 hours after the concrete hardens. However, sawing should be complete prior to the onset of stresses from curling and contraction. A good rule of thumb is to complete sawcutting before the new concrete floor surface cools. Cooling normally occurs the same day as the concrete placement (depending on the time of day that placement began).

Contraction or curling stresses can be increased unexpectedly. Sudden rain-showers falling on exposed fresh concrete surfaces will rapidly cool them. This has the potential to cause restraint stresses and, consequently, random cracking. Protecting the fresh surface from drastic changes in moisture and/or temperature helps prevent unwanted stresses. The HIPERPAV™ computer program addresses strength, stress development, and other conditions related to cracking of concrete pavements. It is a helpful design and troubleshooting tool. By predicting the effect of various environmental factors, potentially harmful (crack-inducing) situations can be avoided. During construction, it can be used to identify protective measures for dealing with unanticipated weather events.

It is important to prevent damage to joint edges. Small, hard wheels can damage the edges of improperly constructed joints. The joints should be filled with a semi-rigid filler flush with the floor surface. If a joint is formed by an insert, the insert should be removed by sawing over the joint alignment and sealing with a semi-rigid filler. Sawing over the strip to remove it creates the filler reservoir. This is the only acceptable way to use plastic strips for forming joints in traffic areas.

Sawcutting the floor to install control joints has traditionally been done by wet sawing with diamond impregnated blades. Special blades are available for sawing hard and soft aggregates. Dry sawing with carborundum blades is done to a limited extent. Dry sawing produces dust, and the sawcut depth must be frequently monitored because of potentially rapid blade wear. A more recent innovation is to cut joints with a relatively small-diameter diamond-impregnated blade. The blade velocity (revolutions per minute [rpm]) is much higher than that of the traditional wet sawcut equipment. Cutting with the high-rpm blade is done considerably earlier than with traditional methods. Depth of cut is restricted to 75 mm (3 in.). Two plates (one on either side of the blade) that rest on top of the concrete surface and confine the concrete mitigate aggregate raveling. The equipment is light enough to use on the surface soon after troweling. If the slab is thicker than about 125 mm (5 in.), it is generally recommended to resaw the crack to the proper depth at a later time before filling with joint

Fig. 7-16. Control joints can be made by a power saw. These joints induce straight-line cracking at predetermined locations. (69657)

sealant. Saw blades should be selected to suit the type of aggregate prevalent locally.

Installation of contraction joints poses a challenge to the surface flatness and levelness characteristics of very flat and superflat floors after slab cooling and drying. When cracks form and widen below the contraction joint notch, upward warping occurs at the slab edges, that is, to either side of the sawcut and below the sawcut crack. The upward warping can be partly mitigated by installing smooth steel load transfer dowels at the control joint.

Tests made within the first 72 hours after concrete placement may not indicate upward slab edge curling at contraction joints. (Traverses for measuring flatness and levelness of floor should neither cross construction nor control joints.) The upward warping due to moisture gradients occurs upon completion of concrete curing when drying begins at the top surface. Using concrete mixes that minimize drying shrinkage can minimize upward slab edge deformations. Shrinkage is minimized for concrete that has:

- low water content
- low paste content
- low slump
- no chloride-containing additives
- maximum coarse aggregate size and amount
- well-graded aggregate

The aggregates used should have low shrinkage potential. Additionally, a proper distribution of all aggregate sizes should be used to minimize voids between aggregate particles and minimize paste demand. The concrete should be placed and cured without causing significant temperature gradients in the slab during the first few days. Good curing practices that minimize evaporation cooling during initial concrete hardening and delay the onset of moisture gradients due to surface drying will help reduce curling. If a small amount of curling does occur, surface grinding can remove the deformation and restore a flat surface at the joint.

Alternately, contraction joints can be minimized by installing narrow lanes between storage racks. Shrinkage-compensating concrete mixes, sufficient reinforcement, and post-tensioned long narrow slabs are various methods of increasing the distance between joints.

Curing

The objective of curing concrete floors on ground is to optimize cement hydration, which is accomplished by maintaining favorable moisture content and temperature conditions in the concrete. With thorough hydration of cement particles, concrete strength development increases and wear resistance of the surface improves. Optimum concrete slab curing conditions range between 10°C and 21°C (50°F and 70°F). At lower than 10°C (50°F), the setting time is extended and finishing takes longer. At higher than 21°C (70°F), the risk of random cracking is greater due to increased warping restraint stresses.

A concrete floor slab has a large exposed surface area in relation to its volume. If the surface is unprotected, water can quickly evaporate and cause early drying. This inhibits cement hydration and can cause a weak concrete surface with poor wear resistance. If the drying is excessive, even light traffic on the slab may result in dusting. Craze cracking is often attributed to inadequate curing. For these reasons, prompt and adequate curing is mandatory. The slab should be continuously moist-cured for at least 7 days (see Figs. 7-17 through 7-20).

For curing, use wet curing or a liquid curing compound. While there are many products that present various claims (multi-purpose liquids), the primary goal of applying a material should be curing. Read manufacturer's literature before choosing a product.

If a moisture-sensitive covering will be applied to the floor, reduce the curing period to 3 days to obtain the best balance between durability and fast drying.

Three alternate methods of curing are suggested:

1. Wet-cure by fully covering the surface with wet burlap as soon as it can be placed without marking the surface. Keep the burlap continuously wet and in place as long as possible. When wetting, the amount of water should be controlled so that no free water reaches the slab bottom. (The magnitude of upward slab edge

Fig. 7-17. A walk-behind curing compound sprayer. (69629)

Fig. 7-18. Water curing is performed by spraying water onto the floor surface. (69688)

Fig. 7-19. Fogging prevents rapid evaporation of water and limits cracking without adding water to the surface. (65114)

Fig. 7-20. Plastic sheets or burlap with plastic backing are also good curing methods. (69668)

warping increases if water is located at the bottom of the slab upon completion of curing.) To ensure continuity of moist curing under burlap without intermittent drying, polyethylene sheets can be laid over the burlap. This will make intermittent wetting of the burlap unnecessary.

2. Wet-cure by fully covering the previously wetted surface with plastic sheeting or waterproof paper as soon as it can be placed without marking the surface, and keep in place as long as possible. This curing method should not be used for slabs to be kept exposed to view because the concrete surface color may be nonuniform. Blotchiness of the surface is accentuated when the polyethylene covered surface is exposed to sunlight during curing.
3. Seal the slab surface and edges by spraying a liquid-membrane-forming curing compound on the finished surface. The curing compound should be a type that leaves no permanent discoloration on the surface and does not interfere with the application of any subsequent surface treatment or overlay. Curing compounds should be avoided on concrete scheduled to receive a floor covering installed with adhesives unless compatibility is established by mockup testing using project materials. Floor areas to be painted, striped, or covered with urethanes or epoxy coatings should not be cured with spray-on sealants unless the slab surface is blasted with an abrasive prior to painting or coating.

Water that is ponded or continually sprayed on the surface is a very thorough method of wet curing, but there are some drawbacks to this type of curing. It limits all other work on the concrete slab surface. Very little can be done until the water is removed when the curing period is over. In addition, the ponded water can seep into joints in the surface or cracks around the perimeter dyke, increasing the water content of the subgrade/subbase. Furthermore, wet curing and wet subgrades or subbases lead to long drying times for the slab, which will delay the application of floor coverings.

A combination of curing methods can be used. For example, a cover of burlap, plastic, or paper can be kept on for 5 days and then a liquid-membrane-forming curing compound can be applied so that the concrete will dry out slowly and extend the curing period.

No matter what type of curing is chosen, joint edges must be cured, too. This is especially critical for joints in industrial floors that will be impacted by hard wheel traffic. If using a liquid curing compound, it should be applied to joint edges. If the curing compound could interfere with adhesion of the joint filler or sealant, the compound must be removed by sawing and grinding. One simple alternative is to fill the joints with moist sand before applying curing compound. The sand should be kept moist during curing.

When moist-curing is completed, drying should begin gradually to minimize shocking the slab. Drying out should be a long, slow process but the drying period can be shortened by special accelerated drying methods. Concrete drying and application of floor coverings are described in detail in Chapter 9.

The surface of a newly completed floor slab should be protected. Subsequent construction activities must not be allowed to damage the surface through neglect and carelessness. These rules should be followed:

1. Keep foot traffic off for 1 day.
2. Keep light, rubber-tired vehicles off for 7 days, unless tests from concrete cylinders cured adjacent to the slab show at least 21-MPa (3000-psi) compressive strength.
3. Leave plastic or waterproof-paper curing sheets in place as long as possible.
4. Protect the surface with sheets of plywood or hardboard where heavy traffic is expected.

There are sure to be changes during construction. Each party should approve of any change. If meetings between the builder and the owner are ongoing, there will be an opportunity to discuss the progress of the construction. Revisions to construction methods during the project can alleviate future problems, may reduce the cost of constructing the floor, and should lead to the best final product.

FLOOR SURFACE TOLERANCES

For a long time, floor surface tolerances were measured by laying a straightedge across the floor surface. The quality of the surface was based on the difference in elevation between the highest and lowest spots over a 3-m (10-ft) length. If the difference was small, the floor was fairly flat in that location; if large, it meant the surface was not as flat. The straightedge would be moved from one location to another to measure floor surface flatness.

In the last decade, other methods for measuring floor surface tolerance were developed because modern vehicles required flatter surfaces for optimum operation, and more equitable standards were needed. Of the best known newer methods, one system is called the F-number system and the other is known as the surface waviness index (Wambold and Antle 1996). These methods offer a significant advantage over the straightedge method because data points are collected and saved. If adjustment is necessary to bring the surface to the required flatness (typically by grinding), the data make it possible to locate specific areas on the floor.

These newer methods look at surface differences over both short lengths and long lengths (chords) and analyze results using statistics. Elevation differences over short lengths tell if the surface is bumpy, wavy, or flat, and over long lengths tell if the surface is sloped, level, or curved. The measurements not only identify these characteristics, but make it possible to quantify the surface, that is, tell *how* flat or *how* level it is. ACI recommends that the floor surface tolerance be measured preferably within 24 hours after placement, but not later than 72 hours after installation. The Canadian Standards Association suggests taking the measurements at 72 hours plus-or-minus 12 hours after completion of the floor finishing.

For the classic method, the 3-m (10-ft) straightedge is placed in any orientation on the floor surface. With the straightedge resting on floor surface highpoints, the depth of a depression in the floor surface is measured. There are four floor classes to describe the slab surface flatness based on the depth of depression. For Class AA, the depression is limited to 3.2 mm (1/8 in.) or smaller; for Class AX, the limit is 4.8 mm (3/16 in.); for Class BX, the limit is 7.9 mm (5/16 in.); and for Class CX, it is 12.7 mm (1/2 in.). As ACI 302 points out, there is no accepted standard test procedure for taking measurements or establishing compliance, so this method of specifying floor surfaces could lead to conflict and litigation.

For the F-number system, also called the "Face Floor Profile Numbers," two numbers—a flatness number (F_F) and a levelness number (F_L) are used to describe the floor surface. With this system, floor characteristics are measured in terms of either metric or inch-pound units, two values for the floor surface are calculated, and results are reported as dimensionless ratings.

The F_F-number provides criteria for short-wave waviness of the floor surface. Measurements are made at 300-mm (1-foot) spacings along straight traverse lines that extend from joint to joint, but do not cross joints. The flatness number quantitatively describes the bumpiness of the floor. The F_L-number provides limits to surface elevation variance from design grade. Measurements are made as for the flatness numbers. The same data used for the flatness numbers are used to generate the levelness numbers, which describe how level or slanted the surface is from one side to another. Notation is always flatness followed by levelness, with a slash separating the two numbers (F_F/F_L).

The F_F/F_L-numbers determine the minimum surface tolerance acceptable for a floor. Measurements to obtain the input data for calculating the F_F- and F_L-number can be made with an optical level, laser level, leveled straightedge, profilograph, dipstick, or digital readout profiler. Measurements and F_F- and F_L-number calculations are made in accordance with ASTM E 1155, *Standard Method for Determining Floor Flatness and Levelness Using the F-Number System.* Though developed in inch-pound units, a complete metric companion to this method has been developed and is designated ASTM E 1155 M. Floor levelness criteria should not be applied to floor surfaces purposely sloped, as, for example, for surface drainage.

The other modern method for assessing a floor surface is the surface waviness index. The waviness index is assessed using the standard method described in ASTM E 1486 M, *Standard Test Method for Determining Floor Tolerances Using Waviness, Wheel Path, and Levelness Criteria [Metric].* (There is a companion inch-pound version designated ASTM E 1486. Solely for purposes of this discussion, metric and inch-pound units appear together. There are two separate standards.) Like the F-number system, it measures differences in elevation over chords of specific lengths: usually 0.6 m, 1.2 m, 1.8 m, 2.4 m, and 3.0 m (2 ft, 4 ft, 6 ft, 8 ft, and 10 ft). Readings are taken along a line to yield a line waviness index (LWI); then these LWIs are combined into a single value, which is known as the surface waviness index, or SWI. This single number describes the degree of planeness of the entire floor surface. This is slightly different from the F-number system, which describes the planeness of the floor with 2 numbers, one for flatness and one for levelness (F_F/F_L). In either system, results are repeatable, but the two systems are separate and distinct. While there is no generally recognized correlation between F-numbers and surface waviness, Table 7-1 provides a comparison of the various flatness measurements—straightedge tolerance, F-numbers, and SWI.

Note, however, that only one system should be chosen to describe the surface characteristics of a floor. Table 7-1 compares all three systems. Supporters of the waviness system, which is often based on chords 0.6 m to 3.0 m (2 ft to 10 ft), claim it is well suited to floors designed for random forklift traffic, because it provides good rideability at typical speeds. The ACI Committee on tolerances is considering the suggested standard tolerances. Vinyl tile covered floors should have an $SWI_{2\text{-}10}$ of about 3 mm (0.12 in.)

as measured by this method. This is roughly equivalent to an F_F/F_L of 28/20 and a 6-mm (1/4-in.) gap under a straightedge. ASTM E 1486 M is the only ASTM specification for narrow aisle, high rack, and defined wheeltrack tolerances.

The range of floor surface tolerances, in general, extends from those with an F_F/F_L designation of 15/13 (SWI about 8 mm) to those that are very flat with F_F/F_L values of 50/30 or higher (SWI about 2 mm or smaller). The sophistication of concrete placing and finishing operations increases with flatness and levelness requirements. This leads to increased costs of placement and finishing. The 15/13 floor criterion can, for example, be met using wet screeds and conventional bullfloats to straighten the surface. The very flat floor, 50/30, demands that side forms be set apart no wider than about 3 m to 8 m (10 ft to 26 ft), that a vibrating screed be used for strikeoff, and that a highway straightedge be used ahead of and between successive machine floating steps. Superflat floor surface tolerances, those greater than 50/50, can be achieved by further reducing distances between forms and additional employment of a straightedge after first troweling. ACI publication 302.1R provides information on strikeoff and finishing techniques to be used for different floor flatness and levelness criteria, as shown here in Tables 7-2 and 7-3.

No matter which method is used to establish floor flatness criteria, surface flatness should be selected on basis of operational performance needs. To minimize construction costs and improve ease of construction, surfaces should not be made more flat or level than necessary. For example, floor areas dedicated for foot traffic with or without carpeting may need no more than F_F/F_L of 20/15 surface characteristics (an SWI of about 5 mm). Floor surface tolerances for offices should have F_F/F_L values of about 30/20 (SWI about 3 mm). Floors surfaced with resilient tile to be kept waxed and polished should be flat to very flat if they are in well-lit areas, because shiny surfaces exaggerate deviations and imperfections. In general, lift-truck traffic surfaces should have F_F/F_L of 20/15 to 30/20 unless otherwise required by a particular rationale or for special equipment.

Lift-truck manufacturers can provide floor flatness and levelness operating criteria for each piece of equipment. Narrow aisles between high storage racks may need very flat (F_F/F_L of 50/30; SWI of 2 mm) or superflat (in excess of 50/30; SWI less than 2 mm) surface characteristics for successful equipment operation—placing and retrieving goods with high mast extensions on lift-trucks. It would not be possible to operate such sensitive equipment like this on a floor that had been finished by floating in a more conventional manner, because the resulting 30/20 tolerance criterion would give unacceptable surface undulations.

A few notable cases of common facilities that have flatness requirements for optimum operation are shown in Table 7-1. These include storage warehouses, television studios, and ice skating rinks.

When high rack storage is combined with narrow aisles, a critical situation exists. Floor flatness and levelness requirements should be made for the longitudinal direction within wheel paths to assure smooth travel as the truck moves along the aisle. The cross-aisle flatness and levelness between the two wheelpaths must also be con-

Table 7-1. Approximate Correlations Between Different Types of Floor Flatness Measurements

Description	Examples	Finishing method	Straightedge tolerance, mm	F_F	F_L	Surface waviness index, mm
Conventional (smooth)	Institutional commercial, industrial floors, both covered and uncovered	Hand screeding with steel trowel finish, no trowel marks or ridges	± 8	20	15	5
Conventional (nonslip)	Base slabs for toppings, exterior flatwork	Hand screeding with broom or float finish	± 12	15	15	8
Moderately flat	Any floor requiring a high degree of flatness and a smooth surface	Highway straightedge (10-m to 15-m wide strips) with steel trowel finish	± 5	30	20	3
			Not recommended	40 to 60	30 to 50	2
Superflat	Movie or television studio	Highway straightedge (narrow strips) with steel trowel ("pizza pan") finish	Not recommended	Over 50	Over 50	Less than 2

Adapted from CSA 1994.

Table 7-2. Slab on Grade—Flatness, F_F

Typical specification requirements		Typical finishing methods
Specified overall value Minimum local value	20 15	1. Smooth surface using 100 mm to 125 mm (4 in. to 5 in.) wide bullfloat. 2. Wait until bleedwater sheen has disappeared. 3. Float surface with one or more passes using a power float (combination or float blades). 4. Multiple passes using a power trowel (combination or trowel blades).
Specified overall value Minimum local value	25 17	1. Smooth and re-straighten surface using 200 mm to 250 mm (8 in. to 10 in.) wide bullfloat. 2. Wait until bleedwater sheen has disappeared. 3. Float surface with one or more passes using a power float (combination or float blades). 4. Re-straighten surface following paste-generating float passes using 250 mm (10 in.) wide highway straightedge. 5. Multiple passes using a power trowel (combination or trowel blades).
Specified overall value Minimum local value	35 24	1. Smooth and re-straighten surface using 200 mm to 250 mm (8 in. to 10 in.) wide bullfloat. Apply in two directions at 45 degree angle to strip. 2. Wait until bleedwater sheen has disappeared. 3. Float surface with one or more passes using a power float (float blades are preferable). 4. Re-straighten surface following paste-generating float passes using 250 mm (10 in.) wide highway straightedge. Use in two directions at 45 degree angle to strip. Use supplementary material to fill low spots. 5. Multiple passes using a power trowel (trowel blades are preferable).
Specified overall value Minimum local value	50 35	1. Smooth and re-straighten surface using 200 mm to 250 mm (8 in. to 10 in.) wide bullfloat or highway straightedge. Apply in two directions at 45 degree angle to strip. 2. Wait until bleedwater sheen has disappeared. 3. Float surface with one or more passes using a power float (float blades are preferable). First float pass should be across width of strip. 4. Re-straighten surface following paste-generating float passes using 250 mm (10 in.) wide highway straightedge. Use in two directions at 45 degree angle to strip. Use supplementary material to fill low spots. 5. Multiple passes using a power trowel (trowel blades are preferable). 6. Re-straighten surface after trowel passes using multiple passes with weighted highway straightedge to scrape high spots. No filling of low spots is done at this stage.

Table 7-3. Slab on Grade—Levelness, F_L

Typical specification requirements		Typical finishing methods
Specified overall value Minimum local value	20 15	1. Smooth surface using 100 mm to 125 mm (4 in. to 5 in.) wide bullfloat. 2. Wait until bleedwater sheen has disappeared. 3. Float surface with one or more passes using a power float (combination or float blades). 4. Multiple passes using a power trowel (combination or trowel blades).
Specified overall value Minimum local value	15 10	1. Set perimeter forms (optical or laser instruments). 2. Block placements of varying dimensions are common. Wet screed strikeoff techniques are used to establish initial grade.
Specified overall value Minimum local value	20 15	1. Set perimeter forms (optical or laser instruments). 2. Block placements of varying dimensions are common. Wet screed or moveable screed strikeoff techniques are used to establish initial grade. 3. Check grade after strikeoff. Repeat strikeoff as necessary.
Specified overall value Minimum local value	25 17	1. Set edge forms using optical or laser instruments. Optical instruments provide more accurate elevation control. 2. Use strip placements with maximum widths of 15 m (50 ft). Utilize edge forms to establish initial grade. 3. Use vibratory screed for initial strikeoff.
Specified overall value Minimum local value	30 20	1. Set edge forms using optical or laser instruments. Optical instruments provide more accurate elevation control. 2. Use strip placements with maximum widths of 9 m (30 ft). Utilize edge forms to establish initial grade. 3. Use vibratory screed for initial strikeoff. 4. Check grade after strikeoff. Repeat strikeoff as necessary. 5. A laser screed can be used in lieu of rigid strikeoff guides and vibratory screed to produce this quality.
Specified overall value Minimum local value	50 35	1. Set edge forms using optical instrument to +/- 1.5 mm (1/16 in.) accuracy. Use straightedge to identify form high spots; plane top surface to fit inside 1.5 mm (1/16 in.) envelope. 2. Use strip placements with maximum widths of 6 m (20 ft). Utilize edge forms to establish initial grade. 3. Use vibratory screed for initial strikeoff. 4. Check grade after strikeoff. Repeat strikeoff as necessary. 5. Follow vibratory screed pass with two or three hand straightedge passes along axis of screed. 6. Use a laser screed in lieu of rigid strikeoff guides and vibratory screed to produce this same quality.

Adapted from CSA 1994.

sidered so that the lift-truck will sit level. For a high-rack storage facility with well-defined pathways (and no anticipated changes in storage rack layout), the very flat floor construction effort can be restricted to narrow, but long, strips.

Movie and television studios require very flat, level floors to enable smooth camera operation. Unlike storage facilities where racks limit travel to specific paths, studios do not have any restrictions for direction of travel of camera dollies. Similarly, in ice skating rinks, the surface of the floor must be very flat and level in every direction, because uniform ice sheet thickness is important to the skating surface. For facilities such as these, special attention to finishing is needed in order to achieve the desired floor profile.

Owners, users, designers, and quality assurance staff are cautioned that requirements for floor surface flatness and levelness agree with anticipated floor uses and not be specified any more restrictive than necessary. Floor installation and compliance costs increase with tighter tolerances.

As previously discussed, specifications require floor surface measurements to be taken within 72 hours after concrete placement. This timeframe provides ongoing control and feedback to the floor installer and allows fine-tuning of the placement, strikeoff, and finishing techniques to meet floor surface tolerances. However, final (ultimate) floor surface levelness, and to a lesser degree surface flatness, can be less than as-constructed and measured (within 72 hours) because following curing, concrete surface drying may lead to upward slab warping. Warping is unpredictable, however, so it is not possible to preshape or camber the slab to compensate for warping. Instead, following drying, remedial measures such as surface grinding may occasionally be required to meet surface operational requirements, even if flatness/levelness criteria were met during construction (see Fig. 7-21).

Fig. 7-21. Grinding can be used to correct slab edges that have warped so that the floor meets specified flatness and levelness criteria. (69685)

CHAPTER 8

TOPPINGS AND FINISHES

Concrete floors are widely used because they are durable and attractive. Leaving concrete floors uncovered in commercial buildings creates opportunities for saving money by:

- eliminating the cost of covering the floor
- creating a low-maintenance, wear-resistant surface
- allowing earlier occupancy
- avoiding potential moisture-related floor covering failures

For a wide range of floor applications, typical concrete provides adequate durability and appearance. In some cases, it is desirable to increase durability or improve the appearance of the floor surface.

One critical measure of floor performance is how well the surface wears. In general, concrete is a hard material that resists wear well. However, a floor surface might lack adequate abrasion resistance even though the floor is structurally sufficient. These floors may require special treatment to create a surface that withstands the abrasive wear loads applied to it, such as heavy equipment traffic.

Abrasion resistance can be improved by applying toppings, aggregate-type surface hardeners, or other special finishes to the concrete surface. Generally, anything that improves concrete strength also improves its resistance to abrasion.

Chemical resistance is another aspect of floor durability. Coating concrete with special materials can prevent chemical attack. These toppings alter the appearance of concrete.

With some floors, appearance is the most important consideration. When concrete is used as a floor finish, the color and texture can be chosen for decorative purposes. Depending on the desired characteristics, the floor appearance may be achieved by use of integral pigments, dry-shake toppings, or surface stains and waxes.

IMPROVING WEAR RESISTANCE

Wear resistance is the ability of the concrete floor to support foot or vehicular traffic without a loss of surface material. One way to classify floors is given by the American Concrete Institute committee on floors (ACI 302) on the basis of traffic exposure and corresponding choices of floor surface and/or floor covering (see Table 8-1). Resistance to wear is classified by degrees: from nominal—that needed to meet demands from foot and pneumatic tire traffic—to extremely high—that needed for special installations, such as waste handling facilities exposed to scraping from tire chains and front end loader buckets. For very high wear resistance floor categories (Classes 7 and 8), special mixes (aggregates) or special finishing techniques (repeated hard-steel troweling), and application of a surface hardener during finishing are design and construction options.

For some facilities, even small amounts of dust due to surface wear can be a problem. This might include processing areas for food, pharmaceuticals, or electronic components. For other facilities, sustaining the structural integrity of the floor is the primary reason for preventing substantial surface wear. This might include any place where heavy equipment processing of bulk materials takes place, such as lumber yards or trash facilities.

Table 8-1. Wear Resistance Rating

Wear resistance	Traffic exposure	ACI 302 Class
Nominal	Foot and pneumatic tires	1, 2, 3, 5
Good	Solid rubber tires	4
High	Hard solid tires, hard casters-light loads	6, 9
Very High	Heavy-duty lift trucks, solid tires, hard casters and steel wheels-medium loads	7, 8
Extremely High	Heavy vehicles with solid wheels, dragging skids, tires with chains and spinup	—

There are a number of pieces of testing equipment that subject hardened concrete samples to abrasive testing. Wear resistance can be measured according to ASTM Designation C 779 using one of the following methods:

- Procedure A, revolving disk
- Procedure B, dressing wheels
- Procedure C, ball bearings

Wear test results differ depending on the test method employed. Revolving disks are used in combination with an abrasive grit and will, with some floor constituents, produce a polished surface similar to terrazzo. Dressing wheels will apply a scarifying wear effect and steel ball bearings will provide some impact.

The three ASTM C 779 procedures differ in the type and degree of abrasive force they impart. They are intended to determine relative variations in surface properties. Results from one test may have no meaningful comparison to results from other tests. This consideration highlights the importance of choosing the appropriate test procedure, one that reflects the type of service (loads) to which the floor will be subjected.

In terms of measuring wear resistance by ASTM C 779 or other methods, depth of tolerable wear and dust formation must be keyed to evaluation of test results that include variables such as:

- concrete compressive strength
- water-cement ratio
- vacuum dewatering
- seeding and incorporating traprock aggregate and grit
- finishing steps and effort
- curing
- liquid-applied surface hardeners
- dry-shake mineral aggregate or metallic surface hardeners
- bonded toppings using special aggregates or materials

Tests made by PCA and the University of California compared four different concrete surfaces (Brinckerhoff 1970). As results shown in Fig. 8-1 indicate, resistance to abrasive wear by the dressing wheel method is less for a 41-MPa (6000-psi) concrete hard-troweled surface than it is for a metallic aggregate surface. Traprock pea-gravel- size coarse aggregate plus traprock grit worked into the surface of the 41-MPa (6000-psi) concrete followed by hard troweling was more resistant than the metallic surface. The best of the four surfaces in terms of wear resistance was the high-strength topping. It was made with traprock and had twice the compressive strength of the other three concretes.

When little or no dusting can be tolerated, the depth of abrasion must be limited to 0.80 mm (1/32 in., or about 30 mils) or less. Based on the dressing-wheel data of Fig. 8-1, there are no significant advantages in using the traprock surface treatment as opposed to the metallic surface treatment. Instead, the high-strength topping made with traprock provides the best wear resistance.

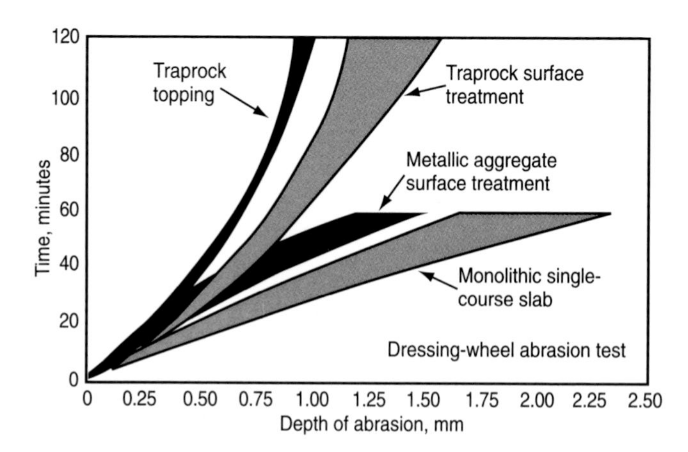

Fig. 8-1. Abrasive wear tests (Brinkerhoff 1970).

The four materials shown in Fig. 8-1 were also abrasion tested by the rotating disk method (ASTM C 779). Analysis of the 30-minute data ranged from a depth of wear of 0.28 mm to 0.56 mm (11 mils to 22 mils). The results were slightly different from those determined by the dressing-wheel method (presented in Fig. 8-1). In order of decreasing wear resistance: metallic aggregate surface was the best, followed by high-strength traprock topping, with very little difference between the traprock-and-grit surface as compared to the normal concrete surface. This clearly illustrates that results differ depending on the test method employed, as noted previously. The test method used to compare the relative topping performance should approximate the exposure (the type of loads anticipated).

Yet another type of abrasion test involves rolling water-agitated steel balls over a concrete surface (Liu 1981 and ASTM C 1138). Fig. 8-2 shows that a decrease in the water-cement ratio of the concrete increases resistance to

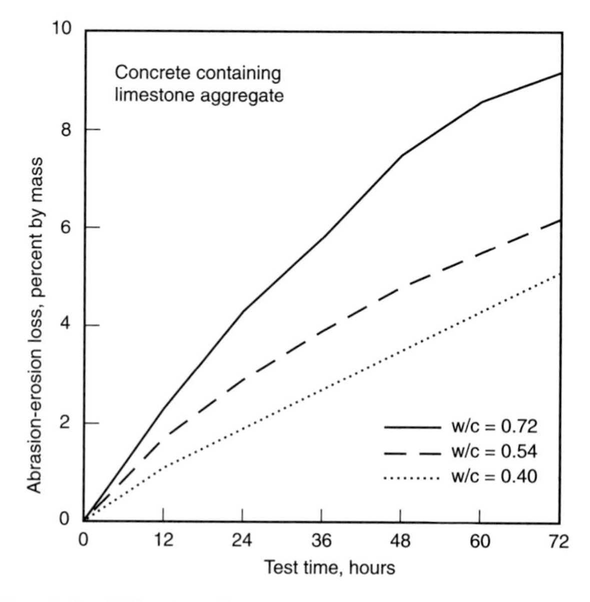

Fig. 8-2. Effects of water-cement ratio on abrasion resistance of concrete containing limestone aggregate (Liu 1981).

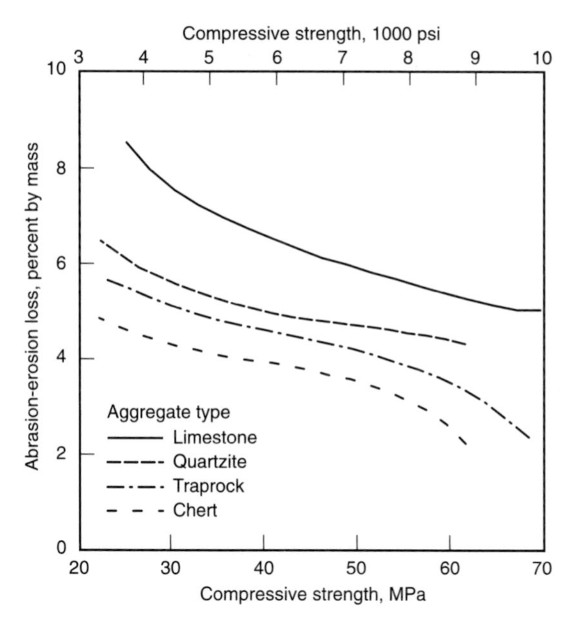

Fig. 8-3. Relationship between abrasion resistance and compressive strength of conventional concrete made with various aggregates (Liu 1981).

wear. Harder coarse aggregate provides greater wear resistance, as shown in Fig. 8-3. Increased compressive strength also increases resistance to wear, regardless of aggregate type.

Based on these and other test results, the following are considered to increase abrasion resistance, both at the surface and at depth in the concrete:

- decreased water-cement ratio
- curing
- increased compressive strength
- increased cement content
- harder aggregates
- repeated power troweling (burnished finish)
- dry-shake surface hardeners, mineral or metallic aggregates
- seeded special aggregates such as traprock aggregate and traprock fines
- high-strength concrete topping
- metallic aggregate topping

Chapter 4 provides detailed information about materials and mixture designs. Consult that chapter for recommended quantities and proportions.

Water-Cement Ratio Reduction

Decreasing the amount of mix water can reduce the water-cement ratio of concrete mixes. However, enough water should be maintained to meet workability considerations. Also, unnecessarily low water-cement ratios on interior floors can result in unnecessary and even undesirable high strength. Water-reducing admixtures are effective in reducing mix water demands. Incorporating hard coarse aggregate coated with cement into the surface will reduce the water-cement ratio of the near-surface mortar. Successive troweling densifies the surface concrete, expelling air and water from voids near the surface, thus reducing the water-cement ratio there.

Vacuum dewatering is a special method of reducing water content in the concrete. Though not commonly used, it is effective when done immediately after concrete has been placed, consolidated, and leveled. The process quickly prepares the surface for final finishing and can reduce drying shrinkage of the concrete, increase compressive strength, and improve the wear resistance of the slab surface.

The basic components of a vacuum system are a plastic-sheet top cover and a bottom filter cloth of nylon netting, which are both placed over the fresh concrete. The top sheet is fitted with suction channels leading to a single vent connected to a vacuum-generating unit. The system can reportedly draw water out of the concrete down to a depth of about 300 mm (12 in.) at an extraction rate of about 3 to 5 minutes for each 25 mm (1 in.) of slab thickness.

Curing

Curing can be done by moist curing methods discussed in Chapter 7. A longer than normal cure period will increase surface hardness and abrasive wear resistance.

Increased Compressive Strength

Decreasing the water-cement ratio will increase the compressive strength. Increased strength can also be obtained by increasing the cement content, though this may also lead to an increase in drying shrinkage. Compressive strength can also be increased by using silica fume, fly ash, or slag in the concrete mix. Strength should be no more than is necessary for load and abrasion resistance.

Increased Cement Content

Increasing the cement content provides for increased strength. Adding cement also creates additional paste, which can be brought to the surface during floating. This results in a thicker surface mortar layer. During troweling, this layer of mortar can be compacted to provide a dense and hard wear-resistant surface. The thicker mortar layer also makes finishing the surface easier and improves embedment of seeded coarse aggregate, when used.

Aggregate Hardness

It is generally accepted that hard aggregates will reduce depth of wear. Tests indicate that troweled surface mortars containing natural sand resist wear better than mortars containing crushed fines (Chaplin 1986). For coarse aggregates, harder aggregates resist abrasion better than softer aggregates, but the effect of hardness is only mobilized after traffic wears away the surface mortar. During normal finishing the coarse aggregate can be depressed to a significant depth below the floor surface. Floors depending on wear resistance from coarse aggregates thus use the mortar above the aggregate as a sacrificial layer.

SPECIAL WEAR RESISTANT FINISHES

Burnished Floor Finishes

A burnished finish is produced by repeatedly power troweling the concrete floor until it has a mirror-like appearance (see Fig. 8-4). Often used on industrial floors to improve wear resistance, the procedure densifies, strengthens, and darkens the surface, while removing pinholes and other surface irregularities.

Fig. 8-4. Often used on industrial floors, a burnished finish can be an economical and attractive alternative to more costly floor treatments for commercial or institutional buildings. (69435)

To burnish a floor requires multiple passes with a trowel. On the first troweling pass, the finisher keeps the trowel blades almost level. On each of the following passes, he increases the tilt angle and rotation speed of the blade.

When a floor has reached the burnished condition, the trowel blades make a ringing sound as they slide over the surface. Because the sound of a trowel blade is subjective, specifiers must make their expectations clear. The best way to achieve a burnished finish may be a combination of a prescriptive and performance specification.

Material requirements include fine aggregate grading limits, minimum cement content, and minimum and maximum slump. Finishing requirements could include repeated steel troweling with a power trowel until the trowel blades make a ringing sound and the floor surface has a glossy appearance. Because "repeated steel troweling," "ringing sound," and "glossy appearance" are all subjective terms, a concrete contractor may be asked to prove his ability to produce an acceptable burnished finish.

Getting an acceptable result is much more likely when the contractor has experience with burnished finishes. Contractors can demonstrate their proficiency by showing the architect examples of burnished floors their firm has placed or by building a mockup of a floor section. Mockups are a good choice because:

- they prove that the contractor has qualified personnel to do the work
- they help identify potential construction difficulties
- the sample becomes a reference/standard of acceptance for the remainder of the floor

As for appearance, surface defects will be more noticeable on a burnished finish than on a surface that is not as smooth, such as a conventionally troweled surface. A burnished finish will not be perfectly uniform, but may exhibit dark colors, craze cracking, circular patterns from power troweling, and other surface blemishes. All of these are cosmetic and do not affect floor performance. It is unrealistic to expect uniform color, but the contractor and concrete supplier can take some steps to avoid discoloration. See Chapter 10.

Burnished floors offer an alternative to conventional finishes, adding value and aesthetic appeal to commercial or institutional buildings. The economics of a burnished floor finish vary from job to job: they often eliminate the need and cost of floor coverings, but increase the cost of finishing compared to a conventionally troweled surface.

Dry-Shake Toppings: Surface Hardeners and Colored Pigments

Dry-shake toppings are sometimes applied to fresh concrete floor surfaces for one of two reasons: to improve the wear resistance or to provide a decorative color treatment. Coloring concrete by the dry-shake method involves the same process discussed below for surface hardeners. Alternately, concrete can be integrally colored with mineral pigments for purely decorative purposes without changing the concrete's abrasion resistance. Additional information about coloring concrete is found in Chapter 11.

Wear resistance can be improved by embedding a select mineral or metallic aggregate into the concrete surface. Materials are sold as proprietary toppings that usually consist of these fine aggregates blended with portland cement. The amount of dry-shake used depends on the severity of anticipated traffic. Other characteristics that affect wear resistance are the aggregate's hardness, gradation, shape, and proportion of aggregate to cement in the dry-shake.

Application of the topping is generally as follows. Light dosages are applied after floating the surface. One-half of the shake quantity is cast over the surface and floated in, followed by partial setting of the concrete (see

Fig. 8-5). When the surface is hard enough to resist a heel imprint, the remainder of the shake is applied and power troweled with steel blades.

For heavy dosages, the application occurs in three stages instead of two. One-third of the shake quantity is spread over the surface immediately after strikeoff. When bleedwater has disappeared, the second one-third of the shake is applied and floated in. After partial hardening, enough to resist a heel imprint, the remaining shake material is floated and power steel troweled to a dense, hard surface.

Shake finishes require special skills and should be applied only by experienced craftspeople. There is a potential for blistering and delaminations when the concrete air content exceeds 3%, and the potential increases with steel troweling. Since dry-shake toppings are steel troweled with many passes, they should not be used with air-entrained concrete mixes.

Colored floors can be achieved by embedding pure mineral oxide pigments or synthetic iron-oxide colorants into the fresh concrete surface. These fine mineral pigments are mixed with portland cement and other materials and intimately blended to assure uniform color results. Because pigmented dry-shakes are used for appearance, select curing products and methods are specifically suggested by the dry-shake manufacturer.

Colors are generally deeper when greater amounts of pigment are used. Though the pigments will adequately stain gray concrete floors, white concrete will provide richer colors. Proprietary products are formulated to match the color of stain with a finishing wax to enhance the color and appearance of the floor surface. Chapter 11 contains information about placing white concrete floors. PCA's *Finishing Concrete Slabs with Color and Texture* (PA124) contains extensive information about coloring concrete surfaces.

Fig. 8-5a,b. Dry-shake is applied in two or three passes to assure complete coverage. (69669, 69670)

Embedment of Abrasion Resistant Aggregates

In some instances, hard aggregates are worked into the floor surface during the finishing process to reduce wear (see Fig. 8-1). Like dry-shake materials, these hard aggregates provide added resistance to wear at the surface, where it is needed. Very hard aggregates are used for this purpose. Aggregate hardness is quantified according to Moh's scale, with higher numbers representing harder aggregates. For instance, diamonds are 10 on the scale and talc is 1. All aggregates fall between these extremes. In Figs. 8-2 and 8-3, Moh's hardness values for limestone, traprock, chert, and quartzite are 3.5, 6.4, 6.6, and 7.0, respectively. These results confirm the beneficial effect of aggregate hardness on resistance to abrasion.

Hard aggregates, such as traprock, are surface wetted then coated with cement; this improves bond. The coated aggregate is broadcast over the fresh concrete surface after strikeoff and floated into the floor surface without excessively depressing the aggregate. Floating and troweling floors with hard aggregate toppings is similar to working floors that have a dry-shake finish (see Dry-Shake Toppings above).

TOPPINGS

High-Strength Concrete Toppings

Where abrasive conditions require concrete of exceptionally high strength, it becomes economical to apply a concrete topping using a different mix. Toppings can be put in place before the base slab hardens (monolithic) or after hardening (bonded).

High-strength concrete topping mixes (monolithic or bonded) should be made with special aggregates selected for hardness, surface texture, and particle shape. Choosing quality aggregate and limiting the aggregate-cement ratio produces the following desirable properties of toppings:

- high strength
- low shrinkage
- abrasion resistance
- good traction (not slippery)

Decreasing the water-cement ratio or increasing the strength of a topping generally improves its resistance to abrasion (see Figs. 8-2 and 8-3). A water-cement ratio of 0.42 or less, an aggregate-cement ratio of 3 (by mass), and good-quality coarse aggregate with a natural sand have proved to be suitable.

Adding silica fume or other pozzolans or slag to the mix can enhance strength and increase abrasion resistance. High-strength mixes can be susceptible to rapid surface drying and plastic-shrinkage cracking. Methods for averting rapid surface evaporation are given on pages 23, 24, 81, and 82. Good curing is essential not only to produce a hard top surface but also to avert warping stresses that could cause debonding before bond strength has developed.

Monolithic Toppings

Monolithic toppings intimately bond to the slab beneath them, so they can be very thin—13 mm to 25 mm (1/2 in. to 1 in.). The base slab should be placed and compacted, then brought to the correct elevation with the topping concrete, which is vibrated while the base is still plastic. In this way the topping becomes part of the structural thickness of the slab. Construction procedures, panel sizes, joints, and jointing arrangements are the same as those specified for the slab itself.

Terrazzo

Portland cement terrazzo is a decorative concrete flooring material that is typically ground and polished to reveal colored aggregate in a white or colored cement matrix. Its hard surface is long wearing, which makes it suitable for heavy duty, high traffic areas where good looks are desired. Some of the common applications are heavily used public buildings, such as airport and train terminals, hospitals, schools, and supermarkets.

The terrazzo surface itself typically is 13 mm (1/2 in.) thick bonded or monolithically installed over slabs. Terrazzo surfaces are separated into panels by metal divider strips that help control unwanted cracking and may also provide decorative patterns, especially when various colors are used on adjacent panel sections. The dividers also serve as leveling guides when placing the terrazzo.

Aggregates most frequently used in conventional ground and polished terrazzo include granite, marble, and onyx chips. The chips range in size from 1.6 mm (1/16 in.) up to 29 mm (1-1/8 in.), but chips can be even wider in breadth for a different appearance. Two parts of aggregate are mixed with one part cement matrix, the material is placed, and additional chips are sprinkled over the topping and troweled into it to achieve the proper consistency and appearance.

After polishing the floor, a coating of clear epoxy sealer can be applied for chemical protection. The National Terrazzo and Mosaic Association provides information about specifying and installing these systems.

Separate Bonded Toppings

Separate bonded toppings 19 mm to 25 mm (3/4 in. to 1 in.) thick are applied after the base structural slab has hardened. A fully bonded topping is considered a part of the structural thickness of the floor because the base slab and the topping act together as a monolithic slab. Being thin, these toppings have little load-carrying capacity of their own and their satisfactory performance depends upon their being fully bonded to the base slab.

In preparation for the topping, the slab must be mechanically roughened so that the coarse aggregate is exposed and the surface blown clean of dust and debris. Preparation can be done by rotomilling followed by sandblasting or by heavy media shotblasting. For base slabs with a thick mortar layer near the surface, however, the scarification need not expose the base slab coarse aggregate. If the base slab concrete is hot and dry, it should be wetted prior to receiving the bonded topping. After wetting, the surface should be damp (no free water).

Immediately before the topping concrete is placed, a bonding grout should be brushed into the surface. One instance when bonding grout may not be needed—which should be demonstrated by testing—is for a fluid topping that contains a high cement content. Grout typically consists of 1:1:1/2 cement-sand-water or neat cement (water plus cement) mixed to a creamy consistency. The slurry is applied in a thin coat (3-mm [1/8-in.]) and covered with a topping before the slurry dries out.

Bond Testing

Good bond between topping and base slab is essential for successful floor performance and high wear resistance. The quality of the bonded installation can be determined by bond pulloff tests using the apparatus shown in Fig. 8-6. To perform the test, a core is drilled through the topping and bond grout layer into the base concrete. A threaded steel plug is epoxy-bonded to the topping surface, and a rod instrumented with a strain gauge is fixed to the threaded portion of the plug (see Fig. 8-7). Tensile force is applied to the rod by means of a hydraulic jack. Bond-strength data are determined from strains in the steel rod at tensile failure.

Examination of the failure surface will demonstrate where tensile failure occurred, which is one of the five following locations:

- within the topping
- within the bond grout layer
- at the interface between the grout and the topping
- at the interface between the grout and the base slab
- within the base slab concrete

A test method described in ASTM C 1404 provides a means to measure the adhesive characteristics of materials used to bond freshly mixed mortar to hardened concrete.

Mode of failure. Because the topping is a high-strength concrete with a low water-cement ratio, its tensile strength is in most instances greater than the base slab concrete. Bonding grouts are also produced with low water-cement ratios, and may also have greater tensile strength than the

Fig. 8-6. Tensile bond pull-off test equipment. (69671)

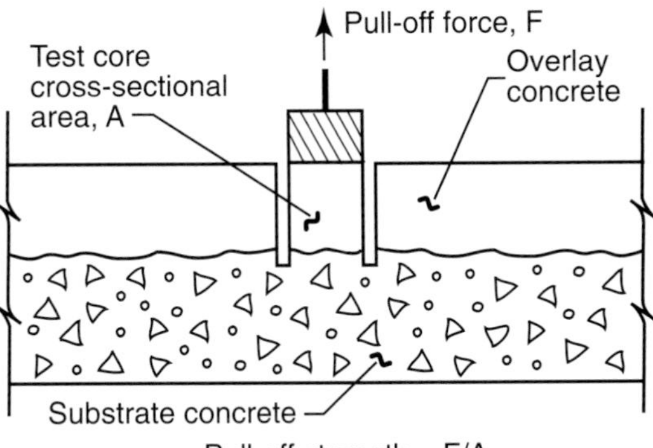

Fig. 8-7a,b,c. Tensile bond pull-off test: performing the pull-off, the resulting core sample, and a schematic depicting the test equipment. (69673, 69674)

base concrete. Most bond test failures will occur within the base slab concrete. When tensile failure occurs at interfaces between grout and the topping or grout and the base, there are a number of probable causes, especially if the tensile strength is less than 1.0 MPa (150 psi). Failure can be due to insufficient cleaning of base slab surface, insufficient effort in brushing the grout mix into the base slab surface, a time delay allowing grout drying ("skinning over" of the grout surface) before placing topping concrete, or sub-standard compaction of the concrete topping.

To determine bond pull-off criteria for toppings, it is essential to verify, by testing, the tensile pull-off value of the base concrete. This is done by coring 10 mm (3/8 in.) into the base slab to create a "plug" where the test apparatus can be attached, then applying tensile force. Acceptance criteria for bonded toppings should not exceed the tensile bond pull-off strength of the base slab.

Joints. *Joints in the bonded topping should only be installed where joints occur in the base slab. No intermediate joints should be installed.* Joints matching base slab joint alignments should be sawcut to full depth of topping. Existing cracks in the base slab should be immobilized by either epoxy injection or by joint stitching. Joints are stitched by drilling vertical holes into the base slab on either side of the crack and installing deformed bars (shaped like large staples) across the crack.

Thickness. Thickness of bonded high-strength concrete should range from about 13 mm to 25 mm (1/2 in. to 1 in.) depending on types and sizes of materials. Thicker bonded toppings make it more difficult to establish good contact with the bond grout layer and to provide good compaction at the underside of the topping. This makes thicker toppings more prone to delaminate at joints and corners due to upward warping of the edge.

Separate Unbonded Toppings

Unbonded toppings are used when the existing floor slab is cracked or otherwise not suitable to receive a fully bonded topping, but is still sound enough to provide a good base for a new slab. The base slab should be covered with a separation sheet of polyethylene or waterproof paper to ensure there will be no bond between the old slab and the new concrete topping. The unbonded topping thickness should be dimensioned on basis of floor thickness design (see Chapter 5), but a minimum topping thickness of 100 mm (4 in.) is recommended.

In the design of an unbonded topping placed over an existing concrete slab, a large value can be used for the modulus of subgrade reaction. In general, 136 MPa/m (500 pci) can be used for interior load positions. However, 54 MPa/m (200 pci) is the maximum subgrade support modulus value given in the nomographs of this publication. This value should be used for the thickness design, and to account for the better support that the base slab will actually provide, the resulting slab thickness determined from these figures can be reduced by about 5%. However, it should be recognized that an unbonded topping slab is prone to severe upward edge and corner warping if it is placed over a polyethylene bond breaking slip-sheet over a very hard base. Thus, this type of construction should be limited to floors that carry foot traffic and, at best, very light carts.

Because there is a bondbreaker between the old and new slabs, joints in the new topping can be laid out in panels to the most suitable size and shape, taking into account the thickness of the topping and the amount of reinforcement provided (if any). Isolation joints in the base slab, however, should be extended up through the topping.

FLOOR HARDENERS

Floor hardeners can upgrade wear resistance, reduce dusting, and improve chemical resistance of a floor surface. These products are normally based on sodium silicate or a type of fluosilicate. It is important to note that none of these products will convert a poor-quality floor into a good-quality floor.

Sodium Silicate (Water Glass)

The cement hydration reaction that takes place during curing produces hydrated lime [$Ca(OH)_2$] within the concrete that is converted to calcium carbonate after prolonged exposure to air. When a solution of sodium metasilicate (Na_2SiO_3), commonly known as water glass, is allowed to soak into the surface, the silicate reacts with calcium compounds to form a hard, glassy substance within the pores of the concrete. This new substance fills the pores and, after drying, gives the concrete a denser, harder surface. The degree of improvement is dependent upon the depth of penetration by the silicate solution. Therefore, the solution is diluted significantly to allow adequate penetration.

The treatment consists of two or more coats applied on successive days. On new concrete, a period of air-drying after completion of moist-curing gives a reasonably dry surface to gain maximum penetration. The floor should be at least 28 days old. The first coat should be a solution of 4 parts by volume of water to 1 part of silicate. The second coat should be the same solution applied after the first one has dried. The third coat should be a 3-to-1 solution applied after the second coat has dried. The treatment is completed as soon as the concrete surface gains a glassy, reflective finish. Water-glass treatment will minimize dusting, improve hardness, and increase resistance to chemical attack, such as from organic acids.

Fluosilicates

Zinc and magnesium fluosilicate sealers are applied in the same manner as water glass. Each of the compounds can be used separately or in combination, but a mixture of 20% zinc and 80% magnesium gives excellent results. For the first application, 3 parts by weight of fluosilicate compound should be dissolved in 25 parts by weight of water. (This is equivalent to 120 g of fluosilicate compound per liter [L] of water, or 1 lb per gallon.) For subsequent coatings, the solution should be 6 parts by weight to each 25 parts of water. (This is the same as 240 g of fluosilicate compound per L of water, or 2 lb per gallon.) To remove encrusted salts, the floor should be mopped with clear water shortly after the preceding application has dried. Safety precautions must be observed when applying these compounds, owing to their toxicity. Fluoridation of concrete surfaces increases the chemical resistance and hardness of the surface.

FLOOR COATINGS

Acids, sugars, and some oils are common chemicals that can attack concrete floors. Spills should always be cleaned up quickly. Coatings are applied to floors to improve chemical resistance, as the concrete by itself resists chemical attack poorly. Coatings also facilitate housekeeping.

The choice of material or coating system should be based on how effectively it resists the anticipated exposure (PCA 2001). When used appropriately, these products offer good abrasion and chemical resistance and can also be pigmented for attractive appearance (PCA 1992). Synthetic floor coatings include epoxies, polymers, polyurethanes, chlorinated rubbers, and phenolics. Many of these are proprietary products and are commercially available.

Common to floor coatings is the need for adequate concrete surface preparation prior to coating application. The manufacturer's directions for concrete surface preparation and application of coatings must be followed closely. New or old, the floor must be treated by shotblasting, sandblasting, or acid etching to remove surface laitance and surface contaminants that could prevent bond between the concrete surface and the coating. Acid etching is used only when other methods are not practical.

In general, the coating installation consists of a primer and the coating. The coatings can be applied by roller, brush, spray, or squeegee. Coating thickness should be as recommended by the manufacturer but must at least exceed surface texture or roughness resulting from surface preparation. Broadcasting silica sand over the wet coating provides traction on the floor surface, especially useful for wet exposures. It should be recognized that coatings, like many floor coverings, are generally impermeable to liquids and moisture vapor. Successful long-term performance depends, in part, on absence of moisture vapor pressures that lead to debonding of the coating. See Chapter 3 for further information on moisture control and vapor retarders.

CHAPTER 9

FLOOR COVERINGS AND DRYING OF CONCRETE

THE IMPORTANCE OF CONCRETE DRYING

Concrete must be allowed to dry before installing a floor covering because moisture affects the installation and performance of floor coverings. There are two main concerns about moisture and its movement: the amount of time needed for concrete drying and the behavior of the floor covering once it is in place.

Excessive floor moisture at the time of installation is one of the main causes of failed floor coverings. Because construction schedules are often designed for the fastest possible completion, floor coverings or coatings may be installed too soon after placing concrete. This increases the risk of failure. Allowing concrete sufficient time to dry reduces the possibility of failed floor coverings. Therefore, it is essential to know the amount of time needed for concrete to dry to an acceptable moisture level. There must also be a way to measure moisture in concrete.

The other concern about floor moisture is how the floor covering performs once it is in place. Although the concrete surface may be dry enough for successful installation when the floor covering is placed, adding an impermeable covering prevents the escape of any additional water vapor that moves to the surface. Ultimately, it is the moisture sensitivity of the floor-covering adhesive in combination with the moisture content of the concrete and the moisture vapor transmission rate through the concrete that determine the performance of floor coverings. Furthermore, water vapor transmission rates can change (increase or decrease) due to changes in the interior building environment or changes outside the building. If vapor collects beneath the floor covering, it can condense, and the water may dissolve the adhesive and lead to a failure (see Fig. 9-1).

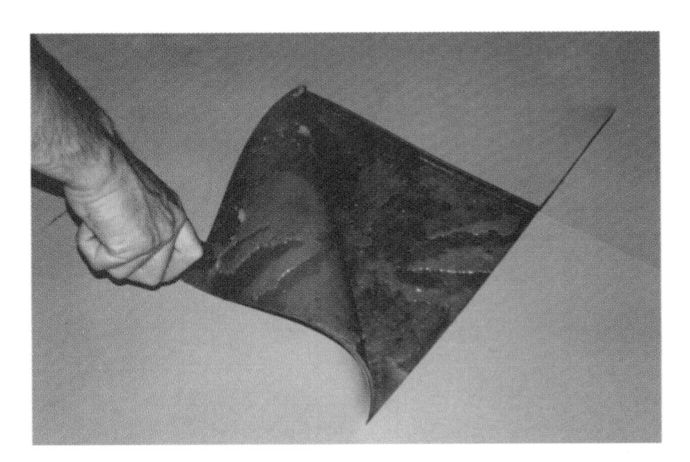

Fig. 9-1. Moisture trapped beneath the floor covering dissolved the adhesive and caused this floor covering to fail. (68129)

In addition to potential floor covering failure, vapor or liquid trapped beneath surface materials creates a moist environment for microbial growth that may produce odors and cause other indoor air-quality problems. Usually, increased amounts of moisture in a building have the potential to lead to greater amounts of damage. The problem can be costly in terms of repair dollars and temporary vacancy during repair.

Concrete Proportioning

Both the water content and water-cement ratio have an effect on concrete drying time. For a fixed cement content, decreasing the quantity of concrete mix water decreases the amount of water that must eventually evaporate. Consistent mix water content can generally be achieved by good slump control. Small variations in mix water content will not significantly affect the drying time needed for floor covering installation.

If a floor is to receive a low-permeability covering, it is beneficial to proportion the concrete mixture with a low water content to speed concrete drying. Some of the water combines chemically with the cement. This hydration reduces the amount of water that can move to the floor surface after curing. Low water-cement ratio concretes undergo more internal drying (called self-desiccation) than high water-cement ratio concretes. Another drawback of high water-cement ratio concretes is their higher capillary permeability, which may allow water from external sources to penetrate the concrete more readily.

A water-cement ratio in the range of 0.40 produces a concrete that has been called "quick drying" (Suprenant 1998a, 1998b). If cured long enough, without adding external water, concrete with a water-cement ratio lower than 0.42 should theoretically use up the surplus water. Concretes that have greater strength *as a result of a lower water-cement ratio* will dry faster than those that have a lower strength and higher water-cement ratio. Otherwise, concrete strength has little effect on the drying time.

Internal drying can also be accelerated by using highly reactive pozzolanic admixtures such as silica fume in the concrete. However, the use of a low water-cement ratio and silica fume generally creates cohesive mixtures. This can make finishing difficult.

Initial Drying

External drying starts when curing is discontinued. This drying occurs by water vapor escaping through capillary pores and microcracks that provide continuous flow paths within the concrete. The capillary pores are formed by remnants of mix water not used for hydration. Microcracks are caused by autogenous shrinkage, drying shrinkage, and thermal contraction. Reducing the water-cement ratio benefits external drying because it reduces the size and number of capillary pores. It also reduces the time needed for some of the capillary pores to become discontinuous and restrict movement of water vapor to the surface. For concrete with a water-cement ratio of 0.7 or greater, capillary pores never become discontinuous, regardless of curing time. Although water vapor moves to the surface faster through these continuous pores, there is more surplus water to be removed. Thus a high water-cement ratio does not help to speed the initial drying process and may cause problems in service if the floor is exposed to an external water source.

High surface air temperature and low relative humidity speed drying. In humid climates, drying takes longer, and may require heat, dehumidification, and ventilation to produce a moisture content that is suitable for floor covering installation.

Beneath the surface, concrete reaches moisture equilibrium very slowly. Fig. 9-2 shows the results of test on a normal weight 28-MPa (4000-psi) concrete that was moist cured for 7 days and air dried at 23°C (73°F) and 50% relative humidity. Concrete 6 mm (1/4 in.) from the drying surface reached 70% relative humidity in 28 days, but concrete 75 mm (3 in.) from the drying surface took 6 months to reach 70% relative humidity. Other studies showed that when 150-mm (6-in.) thick concrete slabs were dried to a target internal relative humidity level of 75%, only concrete that was 19 mm (3/4 in.) or less from the drying surface had a relative humidity gradient (variable relative humidity). Throughout the rest of the slab thickness, relative humidity remained at a fairly constant 75%. This was true whether slabs were dried at an ambient relative humidity of 10%, 35%, or 50% (Abrams and Orals 1965).

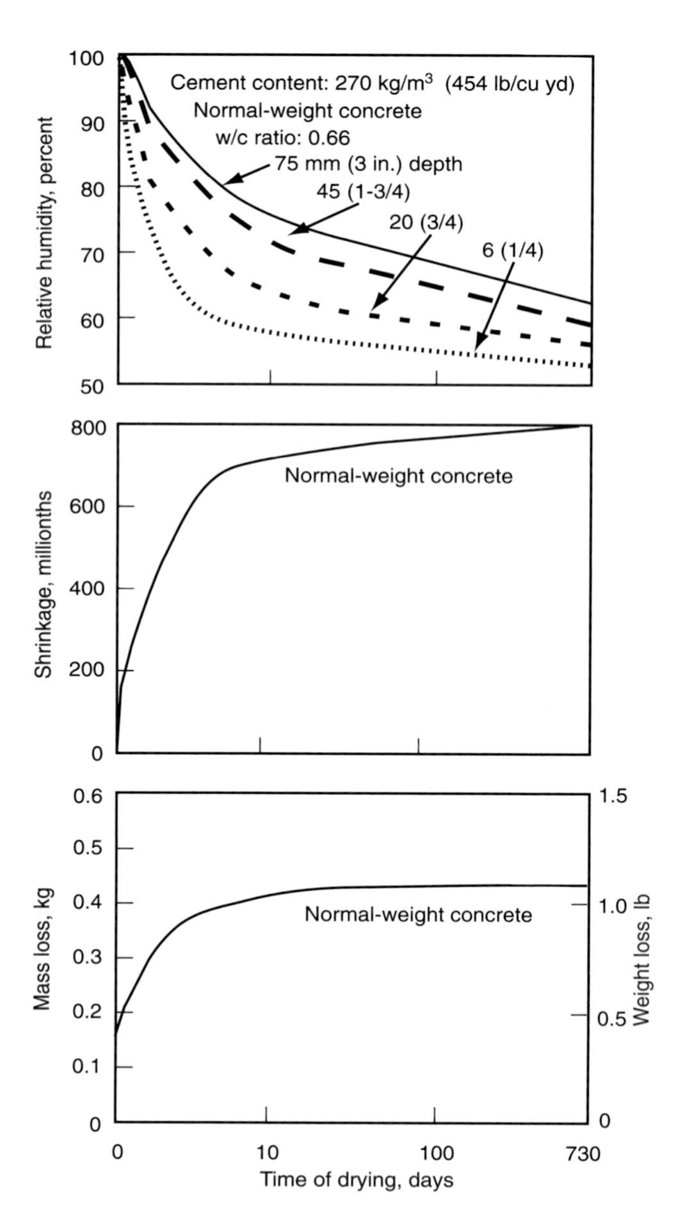

Fig. 9-2. Relative humidity versus depth in 150 mm x 300 mm (6 in. x 12 in.) concrete cylinder. Also shown is mass loss and shrinkage (Hanson 1968).

Moisture Movement in Service

Experience has shown that flooring distress caused by moisture is often not related to the concrete surface moisture content when the floor covering was installed. For instance, in arid climates, the surface may dry quickly while moisture content in the deeper concrete remains high. When an impermeable covering is bonded to the floor, the moisture content beneath it may increase as moisture deeper in the slab moves to the surface.

Changes in external water sources may also cause increased movement of moisture toward the floor surface. A rising water table, heavy rains and poor site drainage, or excessive watering of landscaping are some examples. Changes inside the building can raise moisture content at the concrete surface. During either heating or cooling seasons, lowered relative humidity may draw more water vapor to uncovered surfaces. If the floor is covered with a low-permeability material, temperature changes at the surface or base of the concrete slab may cause water vapor to move either toward the surface or the base.

The best protection against harmful consequences is careful attention to all phases of design, construction, and maintenance of the building.

Water from Within and Below Slab

The presence of water can interfere with floor surface treatments. This is true whether water is in the concrete slab or supplied from moist soils beneath it. Moisture can prevent bond of new coatings or can destroy bond of coatings installed at an earlier time (see Fig. 9-1). Before installing a coating or covering, the concrete floor surface should be tested to determine concrete floor moisture conditions.

Moisture Measurement

Tests for moisture conditions are designed to tell if the floor conditions are acceptable for installing coatings or coverings. Results may be qualitative or quantitative, and are reported in different ways, depending on the method. They can be visual observations, physical tests, or measurements of water quantities or percentages.

The moisture level in concrete can be expressed as relative humidity, moisture content, or moisture ratio. The in-place relative humidity can be determined from a moisture sensor placed in a bore hole drilled into the concrete, from a core drilled from the concrete and calibrated with a sensor, or by an in-situ meter.

In all types of moisture measurement, there can be errors. To minimize errors, only trained professionals should measure moisture in concrete.

Water vapor transmission occurs due to the following conditions, which create a driving force:

- higher relative humidity below the slab
- lower relative humidity above the slab

For uncovered floors, water movement through the slab is normally not a problem, because surface evaporation is more rapid than water vapor transmission.

It is not necessary to have free water beneath a slab for moisture vapor transmission through the floor. Even though the soil appears completely dry prior to placing concrete, there is no assurance that water vapor transmission will not occur through the floor at some point in the future. In fact, altering the facility interior ambient conditions can create a driving force for vapor, drawing moisture through the slab into the building. For example, increasing interior temperature by 6°C (10°F) and reducing ambient interior humidity by 20% (from 80% to 60% due to the drying effect of heating) will about double the vapor pressure differential.

There are many different types of tests and equipment for measuring the moisture condition of the slab:

- polyethylene sheet test
- mat bond test
- gravimetric moisture test
- calcium chloride test
- relative humidity probes
- electrical resistance meters
- electrical impedance test
- nuclear moisture gauge

Consult the floor covering manufacturer about choosing the most appropriate test for their product and what degree of dryness should be attained for successful floor covering installation.

Some of the quantitative tests described here are subject to many interferences and are not recognized by standards setting bodies or by flooring manufacturers for the purpose of accepting or rejecting the moisture content of a concrete floor.

Plastic sheet test. **(ASTM D 4263, *Test Method for Indicating Moisture in Concrete by the Plastic Sheet Method*).** This simple test involves taping a 460-mm (18-in.) square of 0.1-mm (4-mil) polyethylene onto a concrete surface and allowing it to remain for at least 16 hr, then examining the underside of the sheet and the concrete for signs of moisture (see Fig. 9-3). If condensed moisture is present under the sheet, or if the concrete has darkened noticeably, then excessive moisture is present and the concrete is not ready to receive a moisture-sensitive covering. Some flooring manufacturers specify a 24-hr test period using heavy-duty polyethylene sheeting.

Fig. 9-3a,b. The plastic sheet test. (69689, 69690)

This test can be misleading. Although easy to perform, the test provides unreliable results. In order for visible moisture to condense under the plastic sheet, the temperature on the concrete surface must be below the dewpoint. If the (ambient) temperature is warm enough, even a high moisture content in the slab will not cause condensation under the plastic sheet. That result would be termed a false negative.

The plastic sheet test only indicates moisture problems that might occur within the first few days following installation of the floor covering. This would result from moisture near the concrete surface. Problems that might develop over a longer period, such as moisture vapor migration from the subbase into the slab, are not detected by this test.

Mat bond test. In this test, a 1-m (3-ft) square sample of resilient sheet flooring is adhered to the concrete floor using the manufacturer's recommended (written) adhesive and installation procedure. The edges of the flooring are taped to the concrete. (Some instructions suggest applying the floor immediately after applying adhesive, with no open time, to see if the excess moisture from the adhesive remains wet or soaks into the concrete.) After 72 hours the flooring is pulled up by hand. The force required to remove the flooring is judged, and the condition of the adhesive is examined. If the adhesive is re-emulsified or obviously wet, or if the bond is unacceptably weak, then the floor is not ready to receive flooring. Wet adhesive can also indicate an impermeable sealer is present on the surface of the concrete.

This technique requires judgement and experience to evaluate the quality of adhesive bond. A well-bonded sample suggests that the floor is suitable for installation of the flooring.

The mat bond test only indicates moisture problems that might occur within the first few days following installation of the floor covering. This would result from moisture near the concrete surface. Problems that might develop over a longer period, such as moisture vapor migration from the subbase into the slab, are not detected by this test.

Gravimetric moisture content. The weight percent of free moisture in concrete can be determined from a representative sample of a floor slab. The best sample is a full-depth core with diameter at least three times the aggregate topsize. The core should be dry-cut to avoid introducing additional water from the coring operation. Alternatively, pieces of concrete can be cut (with a dry masonry saw) and chiseled from the floor, being sure to go deep enough to represent the bulk of the slab, not just the top surface. The sample must be sealed immediately in a moisture-proof container (or wrapped in impermeable foil) so its moisture content does not change during transport and storage.

In a laboratory, the concrete is weighed and then heated at 105°C (220°F) for 24 hr, cooled in a desiccator, and reweighed. The weight loss is calculated and expressed as percent of the dry weight.

This technique can produce a very accurate measure of the weight percent of free moisture in concrete. *However, free moisture determined by this method does not correlate well with field performance of adhesives and floor coverings.* While gravimetric moisture measurement is an indispensable tool for assessing the moisture content of aggregates, soil, and subbase materials, it is not very useful for assessing the readiness of a concrete floor to receive floor covering. *Water vapor*, not liquid water, is the major cause of moisture problems in concrete floors.

Fig. 9-4. The mat bond test. (69691)

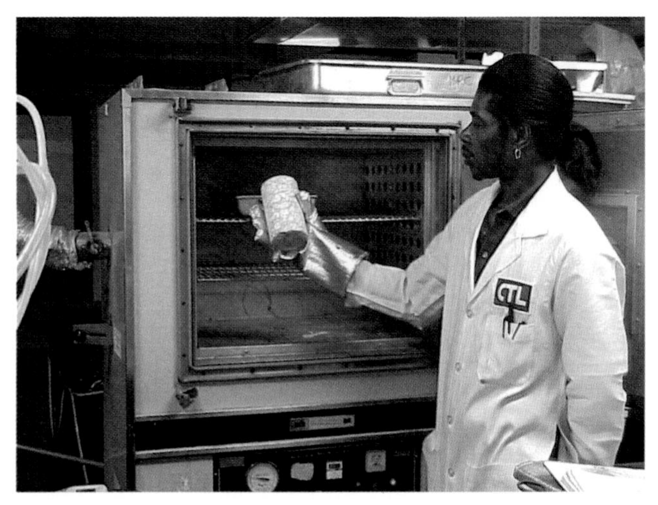

Fig. 9-5. The gravimetric moisture test. (69692)

Fig. 9-6. The calcium chloride test kit. (69693)

Calcium chloride test (ASTM F 1869). (Moisture vapor emission rate, or MVER.) This test is the one most commonly used in the United States and is recommended by the Resilient Floor Covering Institute, the Carpet and Rug Institute, and many floor covering manufacturers.

This test expresses moisture vapor emission from concrete floors as pounds of moisture emitted from 1000 sq ft in 24 hours. Specification limits vary by flooring manufacturer and material type; typical maximum MVER limits are:

5 lb/1000 sq ft/24 hr for: vinyl composition tile, felt-backed resilient sheet flooring, and porous-backed carpet

3 lb/1000 sq ft/24 hr for: solid vinyl sheet flooring, vinyl-backed carpet, and non-porous-backed carpet

MVER test kits consist of:

1. A plastic dish with lid approximately 75-mm (3-in.) in diameter containing 15 g to 30 g (1/2 oz to 1 oz) anhydrous calcium chloride, tape to seal the lid to the floor, a paper label to record data on the top of the lid, and a "dry bag" to contain the dish during storage until needed for use.
2. A flanged, clear plastic cover, (sometimes called the "dome") 30 mm (1.5 in.) in height and encompassing an area of 460 cm^2 (0.5 ft^2).
3. Caulk to form an airtight seal between the dome and the concrete floor.

Because ambient relative humidity and temperature can significantly affect test results, the building must be enclosed with its heating/ventilating/air-conditioning systems operating. After conditioning the room and floor to the anticipated service conditions for at least 48 hours, select test areas to represent the entire floor, including the center and perimeter of the floor, and run the test.

The floor is prepared by scraping or brushing to provide a clean surface. A calcium chloride dish is weighed to the nearest 0.1 g (including the lid, label, and tape). The label is marked with the starting date, time, weight, and test location. The dish is opened, placed on the floor, and covered with the plastic dome. After 72 hours, the dome is cut with a utility knife, the dish removed, and the lid sealed to the dish using the tape. Then the dish is weighed and the net weight gain in grams is calculated. This is the amount of water that the calcium chloride has absorbed during the test period.

Calculate the moisture vapor emission rate in lb/1000 ft^2/24 hr as shown in the kit manufacturer's instructions. Because ambient air humidity and slab temperature can significantly affect the reported MVER, it is useful to measure and report these data along with the MVER results.

The MVER test determines moisture emitted from the upper few centimeters (roughly an inch) of a concrete slab and is not a good indicator of moisture deep in the slab. A typical, 100-mm (4-in.) thick concrete floor slab requires several months to equilibrate with moisture above and below. Therefore, the MVER test yields just a snapshot-in-time of moisture emission from the upper portion of the concrete and cannot predict the long-term performance of a floor, especially if there is no vapor retarder below the slab. As with the qualitative tests discussed previously, a high MVER result indicates a floor is not ready to receive flooring, but a low MVER result only indicates that the moisture level in the upper portion of the concrete may be acceptable at the time of the test.

Relative humidity measurement. In several countries outside the United States, standards for floor moisture are based on relative humidity (RH) within, or in equilibrium with, the concrete floor slab. This practice has several advantages over other concrete moisture measurement techniques:

1. RH probes can be placed at precise depths in a concrete slab to determine the relative humidity below the surface or to determine the RH profile as a function of depth.
2. RH probes actually measure the relative humidity within the slab and are not sensitive to short-term fluctuations in ambient air humidity and temperature above the slab.

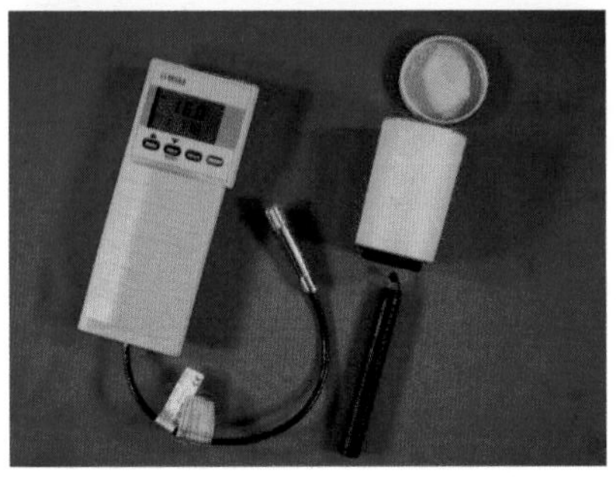

Fig. 9-7a,b. Relative humidity probes. (69694, 69695)

3. Throughout the slab's depth, RH probes directly measure the relative humidity differential, which is the driving force for water vapor movement.
4. RH probes placed at specific depths in the concrete can determine the relative humidity that will exist *after* the floor is covered. When a floor covering is placed on top of a slab it effectively seals the top surface of the slab, or at least significantly reduces evaporation of moisture at the surface; moisture within the slab then distributes itself to achieve a new equilibrium due to temperature and chemical interactions from top to bottom of the slab. Adhesive and flooring are then exposed to the equilibrium moisture level at the top of the slab.

Two British standards, BS5325: 1996, *Code of Practice for Installation of Textile Floor Coverings* and BS8203: 1996, *Code of Practice for Installation of Resilient Floor Coverings,* use a hygrometer or relative humidity probe sealed under an insulated, impermeable box to trap moisture in an air pocket above the floor. The box is sealed to the concrete.

The hygrometer or probe is allowed to equilibrate for at least 72 hours before taking the first reading. Equilibrium is achieved when two consecutive readings at 24 hr intervals agree within the precision of the instrument, generally + 3% RH. Under these two British practices, floors are acceptable for installation of resilient or textile floor coverings when the relative humidity is 75% or less.

The New Zealand Federation of Master Flooring Contractors published a similar method using a hygrometer (Edney Gauge) sealed directly to the concrete floor and covered by an insulated box.

In Sweden and Finland, relative humidity measurements are made by drilling holes in the concrete floor slab and placing probes into the holes (Nordtest Method NT Build 439, *Concrete Hardened: Relative Humidity Measured in Drilled Holes*). For a floor slab drying from its top surface only, a probe placed at 40% of the slab depth (measured from the top of the slab) will determine the relative humidity that will eventually be achieved in the slab at equilibrium after an impermeable floor covering is installed.

RH probes must be set into sleeves that isolate the walls of the drilled holes from the probe so that the probe "sees" only the bottom of the hole. The drilled hole must be allowed to equilibrate for at least 72 hours before making a measurement (it is best to leave a probe in the hole for this period, but the probe can be placed into a previously drilled hole and allowed to re-equilibrate, generally in about an hour). When a hole is first drilled, heat from the friction between the drill bit and the concrete frees moisture that escapes into the hole and nearby concrete; RH measurement made shortly after drilling will yield an inaccurate RH value, which slowly changes over several days to an equilibrium value. Because time, temperature, alkalies, and other factors affect RH measurements in concrete, strict attention must be paid to details of the test procedures.

An advantage of this method is that once a hole is formed in the concrete (or cast into it), the hole can be used repeatedly to check the progress of slab drying. Acceptable RH levels (based on within-slab RH measurements) are given in various sources (see Kanare 2000).

RH probes require regular calibration by the manufacturer (at least once per year). The calibration should be certified and traceable to a national standard (in the U.S. certificates are traceable to the National Institute of Standards and Technology). Users can check performance by placing a probe in a moist room or moist cabinet (protecting the sensor from condensation); after equilibration, the probe should yield a reading of 98%-100% RH.

Another method of determining relative humidity in concrete is to obtain sufficient representative pieces of the concrete from a slab by chisel or hammer drill, place them into a bottle, then cover with a cap having an RH probe sealed through the cap. This method requires that the bottle, RH probe, and concrete pieces come to thermal and moisture equilibrium, a process that typically requires at least overnight. *Do not use concrete powder drilled from a hole for this purpose, since much of the moisture in the concrete is lost due to frictional heat during drilling.*

RH measurements typically are quite precise, + 2% being commonly achieved in the field. However, *accuracy* of RH measurements depends on careful calibration of the sensor and on achieving thermal equilibrium before recording the measurement. At the high relative humidity encountered in new concrete, a safety margin of several percent should be subtracted from the specification limit. For example, if a flooring manufacturer specifies that RH must not exceed 85%, then the RH measured in the field should not exceed 80% - 82% for the floor to be considered ready for installation.

Electrical resistance test. Handheld meters with sensing pins or probes are placed in contact with a surface and the meter is read directly in percent moisture content. Some meters are calibrated especially for wood. Though meters provide results that are accurate (and useful) for wood, they do not work very well with concrete because the surface readings do not correlate with internal concrete moisture. The electrical resistivity of concrete depends on many factors besides moisture content, such as extent of hydration, composition of hydration products, alkalies, and the presence of pozzolans. Pin-type meters only contact the concrete surface and therefore cannot measure the moisture deep within the slab. Such pin-type electrical resistance tests can be quite misleading and are not recommended for any serious floor moisture testing. Some resistance meters use probes placed in holes drilled into the concrete.

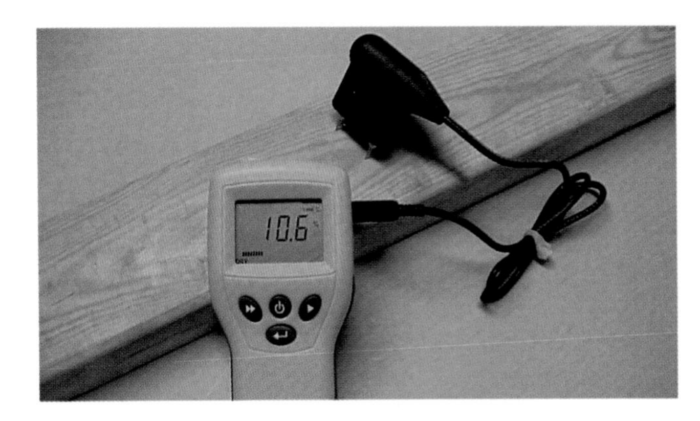

Fig. 9-8. Electrical impedance meter. (69697)

Electrical impedance test. Electrical impedance meters are handheld devices placed on a concrete surface. They measure how electricity travels through concrete. A transmitting electrode on the meter emits a low-frequency electronic signal that is received by another electrode on the meter. The electronic field created by the instrument is affected by the concrete and by moisture in the concrete. Wet concrete is a fair conductor, and dry concrete is a poor conductor. As the moisture level in concrete decreases, the impedance of the concrete increases.

Electrical impedance meters can provide useful information on *relative differences* in moisture conditions to a depth of about 50 mm (2 in.); they are easy to use, provide fast results, and make good survey tools for preliminary testing and for troubleshooting investigations.

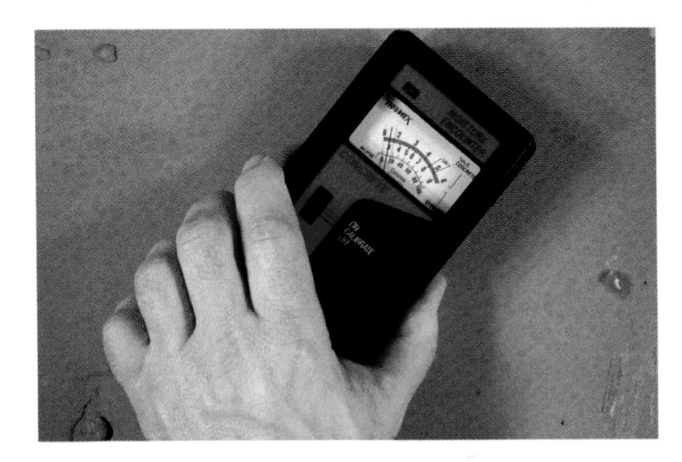

Fig. 9-9. Electrical impedance meter. (69697)

Nuclear moisture gauge. This portable meter contains a radioactive source that emits gamma rays and high-speed neutrons. The neutrons are slowed by interactions with hydrogen atoms in concrete and water, being converted into "thermal" neutrons that are backscattered and detected by a counter in the instrument. A digital display on the instrument indicates the number of counts collected over a fixed time, generally 10–60 seconds per measurement.

A nuclear moisture gauge can read concrete moisture through flooring and floor coatings, but polypropylene-backed flooring can give false high readings due to the concentration of hydrogen in the polymer.

This type of instrument can produce useful information on *relative differences* in moisture conditions to a depth of 100 mm (4 in.). However, this instrument is sensitive to moisture anywhere in the concrete depth, and it does not differentiate between concrete that is wetter at the top or bottom. Concrete has to dry for a long time before this type of meter detects a difference.

Like the electrical impedance meters, the nuclear gauge is easy to use and fast. However, because the instrument contains radioactive material, there are safety precautions that must be taken. Users must be trained and licensed; the owner must be licensed; documents are required to be kept with the instrument, and special fees and travel documents must be obtained for interstate transport; the instrument must be kept locked and placarded with warning signs when not in use.

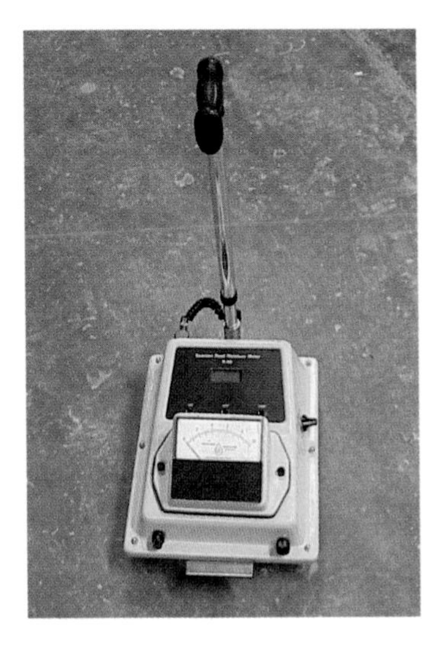

Fig. 9-10. Nuclear moisture gauge. (69699)

The Swedish Method of Estimating Concrete Drying Times

The drying of concrete is complex. It is primarily a function of the amount of mix water and permeability of the concrete. This involves four interrelated factors: water needed for hydration (chemically combined water), water needed for placing, permeability of the concrete, and the moisture gradient.

Drying of concrete, or the removal of uncombined water, starts at the concrete surface and proceeds inward. Measurable factors that affect drying are:

- the water-cement ratio of the concrete
- the thickness of the concrete member
- whether the concrete dries from one or two sides
- the relative humidity (RH) and temperature of the air
- the desired concrete relative humidity (the moisture condition of the concrete at the time a covering is applied)
- the curing method

Drying time. One method of estimating drying time for concrete is based on laboratory research from Sweden's Lund University (Hedenblad 1997). Researchers from Lund's Division of Building Materials predict the drying time (starting *after* concrete curing) of ordinary portland cement concrete using the factors mentioned above. The Swedish approach can help estimate at what age the proper moisture level in the concrete can be reached. This will increase the chances of a successful floor covering installation.

Two pieces of information that must be known are the water-cement ratio of the concrete and the target relative humidity of the concrete. The target relative humidity of the concrete floor is usually based on the type of floor covering and adhesive to be installed; more sensitive coverings require lower relative humidity conditions. Consult the floor covering manufacturer as to the appropriate relative humidity for specific products. From the water-cement ratio and the target relative humidity, a basic estimate of the number of drying days is given. As Table 9-1 shows, it takes approximately twice as long to reach 85% relative humidity as it will to reach 90% relative humidity. These values are modified by the correction factors shown in Tables 9-2 through 9-5.

Member thickness. As members become thicker, the drying time increases. The water-cement ratio has a much smaller effect, but higher water-cement ratios usually lead to a slight increase in drying time as shown by correction factors in Table 9-2.

Table 9-1. Drying Times for Concrete

Relative humidity of the concrete,%	Water-cement ratio			
	0.4	0.5	0.6	0.7
85	50 days	90 days	135 days	180 days
90	20 days	45 days	65 days	95 days

Table 9-2. Correction Factors for Thickness

Thickness, mm (in.)	Water-cement ratio			
	0.4	0.5	0.6	0.7
100 (4)	0.4	0.4	0.4	0.4
150 (6)	0.8	0.8	0.8	0.7
180 (7)	1.0	1.0	1.0	1.0
200 (8)	1.1	1.1	1.1	1.2
250 (10)	1.3	1.4	1.5	1.8

Type of drying. Drying can occur from one or both sides of a concrete member: faster drying occurs when it takes place from two sides (see Table 9-3). Typical concrete members are slabs and walls. Except for basements, walls usually dry from both sides. Floors can dry from one side (top) or from both sides (top and bottom) depending on the subbase material on which the concrete is placed. Suspended structural slabs, as used in multistory buildings, have two-sided drying. Concrete slabs placed directly on the ground, on a vapor retarder, or on a metal deck have one-sided drying.

Table 9-3. Correction Factors for Type of Drying (One- or Two-Sided)

Drying from:	Water-cement ratio			
	0.4	0.5	0.6	0.7
One side	2.0	2.3	2.6	3.2
Two sides	1.0	1.0	1.0	1.0

Drying conditions. Drying conditions are a combination of relative humidity and air temperature. The correction factors for drying conditions are provided in Table 9-4. Warm, dry air dries concrete faster than cool, humid air. In general, the effect of air temperature is more pronounced than the level of moisture in the air, because cooler air does not dry concrete as well as warm air at any relative humidity.

Table 9-4. Correction Factors for Climate (Relative Humidity and Temperature)

Relative humidity of the air,%	Air temperature, °C (°F)			
	10 (50)	18 (64)	25 (77)	30 (86)
50	1.2	0.9	0.7	0.6
60	1.3	1.0	0.8	0.7
70	1.4	1.1	0.8	0.7
80	1.7	1.2	1.0	0.9

Table 9-5. Correction Factors for Curing

Curing conditions	Water-cement ratio					
	0.5		0.6		0.7	
Drying concrete to relative humidity of:	85%	90%	85%	90%	85%	90%
Short curing period	1.0	0.5	1.0	0.5	1.0	0.7
Two weeks rain, two weeks moist air	1.0	1.0	1.0	1.0	1.0	1.0
Four weeks moist air	1.0	0.5	1.0	0.7	1.0	0.8
Four weeks rain	1.4	1.0	1.4	1.3	1.4	1.3

Type of curing. Concretes that are to receive floor coverings will dry faster if they are exposed only to moist air (up to four weeks) or if they receive very little curing (one day in moist air). Wet curing, simulating two to four weeks of rain, adds time to the drying period for the three water-cement ratio levels tested as shown in Table 9-5.

Estimating drying time. The base drying time (from Table 9-1) is multiplied by the correction factors, yielding the number of days required to reach the desired relative humidity of the concrete so that a floor covering can be successfully applied. The following example demonstrates the ease of use of the Swedish method.

Concrete:
- water-cement ratio = 0.6
- 100 mm (4 in.) thick
- placed on vapor retarder (impervious base)
- need 85% RH before covering

Curing conditions:
- 4 weeks moist air

Climate:
- air at 70% RH at 25°C (77°F)

With 85% RH and water-cement ratio = 0.6, the base drying time is 135 days (from Table 9-1).

The correction for the thickness is 0.4 (from Table 9-2).

One-sided drying yields a factor of 2.6 (from Table 9-3).

The combined temperature and humidity give a factor of 0.8 (from Table 9-4).

The factor is 1.0 for curing (from Table 9-5).

The amount of time needed to dry this slab would be:

135 x 0.4 x 2.6 x 0.8 x 1.0 = 112 days, or about 16 weeks

Silica fume can be used as an ingredient in the mix to reduce the drying time: 5% addition of silica fume leads to about a 50% reduction in drying time for a water-cement ratio of 0.5 or lower. Higher doses of silica fume can lead to greater reductions in drying time.

FLOOR COVERING MATERIALS

Table 9-6 provides a list of the vulnerability to loss of bond for some common floor treatments. Note that coatings, coverings, and adhesives that tend to be "impermeable" (or less permeable than concrete) to water vapors are vulnerable to moisture-related debonding or loss of adhesion. The debonding occurs when the adhesive re-emulsifies and loses its holding power. This has happened with quarry or ceramic tiles installed with mastic adhesives. Coverings that are permeable to water vapor movement (cementitious terrazzo, concrete toppings bonded with a cement grout) are not vulnerable.

For new construction, installation of an effective vapor retarder—one that is not pierced during floor construction stages—is inexpensive insurance. Therefore, vapor retarders should be installed below all floor areas scheduled to receive coatings or coverings.

Table 9-6. Moisture Vulnerability of Floor Coverings

Material	Vulnerable	Not Vulnerable
Acrylic terrazzo	✔	
Brick		✔
Carpet with polymer backing	✔	
Ceramic tile		✔
Concrete topping		✔
Epoxy paints and coatings	✔	
Epoxy mortars and concrete	✔	
Epoxy adhesives	✔	
Linoleum	✔	
Linoleum adhesives	✔	
Polyurethane coating	✔	
Polyester coating	✔	
Quarry tile		✔
Resilient tiles	✔	
Resilient tile adhesives	✔	
Rubber floor covering	✔	
Terrazzo (cementitious)		✔
Vinyl sheet goods	✔	
Vinyl adhesives	✔	
Wood flooring	✔	

FLOOR COVERING INSTALLATION

Flexible floor coverings that are attached with adhesives—resilient tiles and linoleum sheet goods—should be installed on top of concrete floors that received hard trowel finishing to a smooth surface. Otherwise, surface imperfections such as small pieces of aggregate raised above the surface could be reflected as bumps in the resilient material surface. The floor must not be cured with an applied curing compound unless tests have shown that the curing compound is compatible with the surface adhesive. The glue is applied to the clean concrete surface with a notched trowel, and the covering is pressed into place. Traffic should not be allowed on the surface until the adhesive has set.

Quarry tile, terrazzo tile, and concrete toppings are installed over a concrete surface prepared by shotblasting or sandblasting. If bonded topping installation is scheduled, the surface preparation should include rotomilling followed by sandblasting. Even newly placed concrete requires shotblasting or sandblasting.

The bond grout for quarry tile can be either a proprietary thinset mixture or a cement mixture. The thinset mixes generally have a latex additive in the mix (usually contained in the liquid portion of the two-component system). The thinset must not be allowed to dry (skin over) before tiles are pressed into place. Neat cement grouts (cement plus water mixed to a thick, creamy consistency) or sand-cement grouts are also used. The thinset or grout is placed on the prepared, cleaned, and dustfree concrete surface with a notched trowel. Tile should make intimate contact with the grout so that its support is uniform, thus averting tile cracking under traffic.

Cast-in-place terrazzo should be installed in accordance with the Mosaic Terrazzo Institute methods (NTMA 1999). Base slab surface preparation requirements are as discussed above for quarry tile installation.

Contraction joints and isolation joints in quarry tile, ceramic tiles, or terrazzo tile coverings should match those in the floor base slab. The joints should be full depth and where exposed to traffic—including foot traffic—they should be sealed with a semi-rigid epoxy flush with the tile surface. Spaces between tiles not aligned over contraction joints should be filled to full depth with a cementitious joint mortar flush with the surface of adjacent tiles. Construction joints in base slabs should be treated as contraction joints in the covering. Isolation joints should be filled with a flexible sealant.

COATING FAILURES

When moisture is present in or beneath a concrete slab, many types of coatings will fail unless they can develop a good bond soon after application. Nonbreathable coatings like epoxy, polyester, polyurethane, vinyl ester, and methacrylate can trap moisture or vapor rising through the slab, leading to blistering or loss of bond. Coatings can appear well-bonded even if they have only developed a bond strength of 0.007 MPa or 0.014 MPa (1 psi or 2 psi). A well-bonded coating, however, will develop a bond strength of at least 0.7-MPa (100 psi), and may even exceed 1.4 MPa (200 psi). Neither a hydrostatic head, capillary action, nor water vapor exerts enough pressure to dislodge a coating that has a 0.7 MPa (100-psi) bond pull-off strength. In other words, if a coating can develop good bond at the time of application, and if the glue is resistant to moisture (does not dissolve over time), moisture that rises through the slab will not lead to loss of bond with the coating. Instead, coatings that suffer moisture-related damage probably never achieved a good bond to the concrete (Gaul 1996).

Moisture can cause coating failures in two ways: by preventing proper cure, or by allowing cure but preventing good bond. For some coatings, the chemical cure is interrupted or severely retarded by the presence of moisture. This results in a soft coating or one that is sticky on its underside. There may be a noticeable odor from one of the coating components. Visually, small blisters (diameter less than 6 mm [1/4 in.]), closely spaced, will be noticeable.

If the coating cures properly and completely in the presence of water, there is still the possibility that it will not achieve good bond under certain moisture conditions. The experience of the coatings industry has shown that a moisture vapor transmission rate of more than 2.5 kg per 100 m^2 per 24 hr (5 lb per 1000 ft^2 per 24 hr) (as measured by ASTM F 1869) can lead to failure. If this type of failure occurs, the coating has a hard underside that can be easily peeled from the surface using a blade (putty knife). Whereas coatings whose cure is *prevented* by moisture exhibit small blisters, *poorly-bonded* coatings exhibit larger diameter blisters. If osmotic pressure develops, these blisters may contain water under pressure, which will squirt out when the blister is punctured.

Before a coating is applied, the moisture condition of the floor must be known. See "Moisture Measurement" earlier in this chapter for a description of various methods available for checking moisture in concrete.

To avoid moisture-related failures of floor coverings due to poor bond, several approaches are suggested. A vapor retarder located under the slab is the first line of defense. This prevents excess water from reaching the underside of the slab.

Breathable coatings. When possible, a coating that breathes should be used. The trouble with these coatings is that they are not generally very resistant to chemicals, do not protect the concrete from water penetration, and are more difficult to clean.

Proper ambient conditions. Turning off heating or air conditioning systems reduces the moisture gradient between the underside and top of a slab. Concrete should be protected from the sun when coatings are applied. If concrete is warmed by the sun, water vapor pressure in the capillaries increases, and a vapor pressure gradient is established. This tends to push coatings off the surface. Installing the coatings in the shade and when the concrete temperature is dropping helps to pull the coating into the concrete.

Primers. A fast-setting primer will bond quickly to the slab and should be considered. This material can develop bond quickly, before moisture collects under the coating or before vapor pressure builds up to interfere with the bond.

It may be beneficial to install a fast-setting primer to any concrete slab that is to be coated.

Intermediate toppings. Use of a semipermeable, breathable membrane can reduce the vapor transmission rate to less than the 1.5 kg per 100 m^2 (3 lb per 1000 ft^2). Materials for this purpose include latex-modified cementitious coating applied at a thickness of 1.5 mm to 3 mm (1/16 in. to 1/8 in.). A fast-setting primer should be applied over the semipermeable membrane.

Alkalies. Ordinary portland cement concrete is highly alkaline, having a pH greater than 12.5. In hardened concrete, the high pH is beneficial because it inhibits corrosion of embedded steel, such as reinforcing bars and wire mesh.

However, when carbon dioxide in ambient air reacts with concrete, it causes a thin layer of carbonated cement paste to form on the concrete surface and reduces the pH to close to neutral. This reaction is known as carbonation. Typically, carbonation proceeds inward from a concrete surface at a rate of roughly 1 mm per year, depending on relative humidity and temperature. It occurs most quickly at about 50% relative humidity.

Because alkalies from concrete can attack adhesives and floor coverings, a small amount of surface carbonation is desirable for concrete floors. Floor covering problems can be the result of:

- flooring installed before the concrete has carbonated sufficiently
- moisture/vapor carrying dissolved alkalies to the surface while traveling through the concrete

CHAPTER 10

PROBLEMS, MAINTENANCE, AND REPAIR

Concrete floors are regularly installed without complications. It is to the credit of concrete that so few complaints are received about slab construction. While it is easy to determine the properties of the hardened concrete that will be suitable for the intended purpose, achieving these properties requires care, as they are affected by each step in the design and construction process. Adhering to proper construction methods maximizes desirable properties and minimizes blemishes.

SURFACE DEFECTS AND PROBLEMS DURING CONSTRUCTION

Many surface defects can be traced to problems during finishing and curing. Therefore, any discussion of surface defects should include a discussion of potential construction problems. Knowing the causes of surface blemishes suggests preventive measures. Common topics include:

- excessive bleeding
- concrete setting time delay
- blistering
- delamination of surface mortar
- plastic shrinkage cracking
- crazing
- carbonation and dusting
- popouts
- random cracking
- discoloration

Bleeding and Set Retarding

Excessive bleeding that occurs after concrete placing, strikeoff, and bullfloating can delay subsequent finishing steps. In most instances, the cause of excessive bleeding is due to one of the following:

- a water-cement ratio that is too high
- poor aggregate gradation
- slow set times
- ambient conditions that hinder surface water evaporation: low temperatures, high humidity, or lack of air movement

Fig. 10-1. Heavy bleeding on a concrete surface. (P29992)

There are a number of ways to handle excessive bleedwater on a fresh concrete surface (see Fig. 10-1). Water collecting on the slab can be removed by dragging a squeegee or hose over the surface. Water can also be removed by blotting with burlap. Cement should not be placed directly on the fresh concrete surface to soak up excess bleedwater. Another way to (promote evaporation and) reduce the amount of bleedwater on the fresh surface is to circulate warm air from vented heaters.

Altering the mix proportions can reduce bleeding. Increasing the amount of cement (or other fines) in the mix allows all particles to remain better suspended. Although small amounts of entrained air may reduce bleeding, in general, the air content for hard troweled surfaces should not exceed 3%, as this can contribute to blistering and surface mortar delamination.

Concrete setting time can be delayed by cooler-than-anticipated temperatures and by admixture incompatibility. When finishing delays do occur, excessive manipulation of the concrete should be avoided, since it could lead to a weak surface. Instead, it is best to wait until setting has occurred or to find a way to decrease the time needed for setting. There are a number of options to speed the set of concrete. Faster set can be achieved with materials—cements or admixtures—or by adding heat.

Increasing the cement factor speeds up the set. High early strength cements like Types III and HE are formulated to react faster than other cement types. Admixtures that speed setting are called accelerators. Calcium chloride is the most common accelerator. It is added with the mix water. However, it should be avoided whenever concrete contains reinforcement. Also, calcium chloride increases concrete volumetric shrinkage and may discolor concrete. Admixtures without chlorides can also be used to accelerate concrete hardening.

Another way to decrease setting time is to increase concrete temperature. This can be achieved by heating the mix water or by placing concrete on a warmed subgrade. Heating the subbase/subgrade works best for interior locations (protected by walls and a roof). It should be started several days ahead of concrete placement. Heaters should be vented to the building exterior to prevent early carbonation of fresh concrete. Otherwise, a soft floor surface that is prone to abrasive wear and dusting could result.

Fig. 10-2. Blistering can be corrected if caught at the proper time. (Inset 49411, primary A5272)

Fig. 10-3. Delaminations affect a larger area than blistering. (67196)

Blistering and Delaminations

Blistering is the convex raising of the surface mortar layer while the concrete is still plastic (see Fig. 10-2). The blisters are attributed to sealing the floor surface before all the bleedwater and air have escaped. With some concrete mixes, sealing occurs as a result of bullfloating or premature floating. If noticed prior to concrete set, the surface can be opened by use of a wood float so that bleed air and bleedwater entrapped in blisters can escape to the surface. If not noticed before concrete hardening, the water or air will remain trapped beneath the surface. Blisters are usually about 50 mm (2 in.) or less in diameter and can easily be seen if lit from behind (using an oblique light source at the opposite side from the viewer's location). Blisters are crushed under traffic and leave about 3-mm (1/8-in.) deep pockmarks.

Similar to blistering, delamination of surface mortar can occur due to entrapment of bleedwater and air below the prematurely sealed mortar surface (see Fig. 10-3). Delaminations affect larger surface areas than blisters, and are very difficult to detect during finishing. They become apparent after concrete surface drying when the delaminated area is crushed under traffic. The thickness of delaminated mortar ranges from about 3 mm to 9 mm (1/8 in. to 3/8 in.). The affected area can be anywhere from a few square centimeters (inches) to a few square meters (yards).

The potential for delaminations increases as the number of risk factors increases. The risk factors are anything that affects the bleeding and setting of the concrete. These include:

- cold subgrade
- vapor retarder directly under the slab
- thick slab
- high water content
- entrained air, total air content
- some chemical admixtures
- some fly ashes and slags
- high wind velocity
- low relative humidity
- direct sunlight and rising air temperature

When one or more of these factors are present, correct finishing techniques become even more important. There are often multiple solutions to potential blistering/delamination problems. Prevention should take the simplest route. Some conditions are easily adjusted, while others are not. For instance, it would be easy to warm the subgrade so the entire thickness of concrete sets at approximately the same rate as the surface, but it would be difficult to change the slab thickness, which is dictated by slab structural design.

If the concrete has stiffened from the top down, as it often does when wind speeds are higher, there is a tendency to finish the slab too soon, before bleeding is complete. Finishing operations performed while the underlying concrete is still soft (and bleeding) will seal the slab surface, potentially trapping bleedwater and leading to delaminations.

Plastic Shrinkage Cracking

Plastic shrinkage cracking is due to concrete at the surface drying (and shrinking) before initial set of the concrete occurs (see Fig. 10-4). Plastic shrinkage occurs during and after finishing, usually when there is rapid evaporation of bleed water. The conditions that lead to rapid water evaporation are low relative humidity, high air temperatures, rapid air movement (wind) across the concrete surface, and elevated concrete temperatures. Under these conditions the concrete surface can crust over while the underlying concrete is still plastic. As plastic shrinkage cracks form, they start at the surface and may extend some depth into the unhardened concrete. Floating the concrete slab can repair plastic shrinkage cracks, but only if done immediately as the cracks occur. Protective measures should be used to prevent formation of the cracks. These measures include:

- fog sprays that raise the relative humidity in area above the floor
- sun shades
- wind breaks
- reducing concrete mix temperature as delivered
  - by replacing a portion of the mix water with ice
  - by cooling the aggregate piles
  - by adding liquid nitrogen to concrete in the ready mix truck
- plastic fibers

When fog sprays are used, mist water should not collect on the concrete, as this could increase the water-cement ratio of the surface mortar and cause a weak floor surface.

Evaporation-reducing chemicals can be applied to the concrete to prevent rapid surface water and bleedwater evaporation during finishing. These products are not curing compounds, but form monomolecular films that decrease the evaporation rate in order to control plastic shrinkage cracking. They should be used according to manufacturer directions. They are not to be used during final troweling because they could lead to surface discoloration.

Crazing

Craze cracks are fine random cracks or fissures in a concrete surface (see Fig. 10-5). On concrete flatwork, they usually extend less than 3 mm (1/4 in.) below the surface. The cracks occur within the paste-rich surface mortar and generally pass through the paste and not through aggregate particles. It is typical for the cracks to form a map pattern. The narrow cracks are so fine that they are difficult to see. In many instances, they are only visible during the drying phase of a wetted surface or when a translucent coating is installed. Craze cracks are attributed to inadequate curing that leads to concrete surface drying and cooling before the mortar has gained sufficient strength. These are cosmetic blemishes that generally have no effect on the serviceability or durability of the floor.

Carbonation and Dusting

Carbonation of a fresh concrete surface occurs when carbon dioxide combines with calcium hydroxide in the concrete to form a layer of calcium carbonate on the surface. This layer is weak and will most likely dust under traffic. Carbonation is more of a concern when heaters are used to protect the fresh concrete during cold weather; if the heaters are not vented to outside air, carbon dioxide from the exhaust can accumulate over the surface of the fresh concrete.

Fig. 10-4. Plastic shrinkage cracking on a surface. (1311)

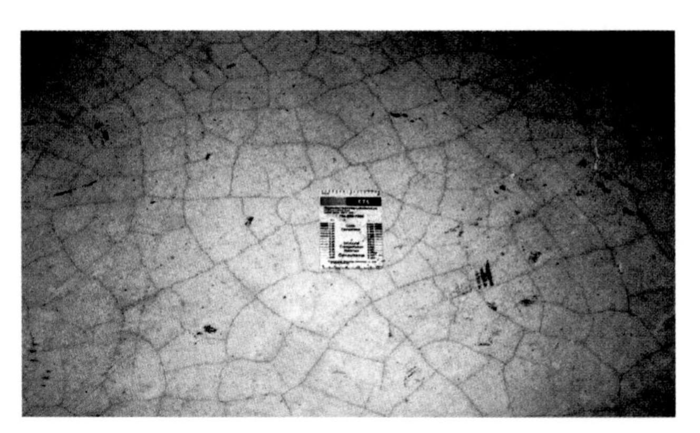

Fig. 10-5. These crazing cracks appear more visible because a clear urethane sealer has been applied to the surface. (69675)

There are other causes of dusting surfaces. Like carbonation, any of the following causes can lead to weak surfaces and dusting:

- water applied during finishing
- exposure to rainfall during finishing
- spreading dry cement over the surface to accelerate finishing
- a low cement content of the concrete
- a mix that is too wet
- too little curing or improper curing (if the concrete dries out too fast)
- freezing of the surface
- dirty aggregate

To improve a concrete surface that exhibits dusting, a surface hardener can be applied as discussed in Chapter 8, Toppings and Finishes, under the section on floor hardeners. Another method to remedy a dusting surface is to grind the top layer of the floor to expose sound underlying concrete, though this will change the surface appearance.

Curling

When the edges and corners of a floor slab on ground dish upward in the absence of any loads other than gravity, the slab is said to be curling. It is usually attributed to differences in moisture content or temperature from top to bottom within the slab. These temperature and moisture gradients develop between the top and bottom surfaces as the concrete in a floor slab hardens. The slab will curl up if the top is drying and cooling (shortening) while the bottom remains moist and warm. Under opposite conditions, the slab should theoretically curl down. Downward curl as such, however, does not occur due to subbase restraint.

Ideally, a slab without temperature or moisture gradients lies flat on the subgrade with uniform support. But, at an early age, concrete near the top of the slab loses moisture due to early drying shrinkage while the bottom retains moisture. As long as the subgrade continues to subject the bottom of the slab to an atmosphere more humid than the top, the curled shape will persist.

This change in slab shape causes the corners and edges of panels to lift away from the subgrade. This not only reduces the support given a floor by the subgrade, but also increases the flexural stresses and deflections induced by wheel loads positioned at the corners and edges of panels.

The tendency to curl is resisted by the weight of the floor slab itself acting to hold the slab in its original position. This restraint results in curling stresses that increase from the periphery of a floor panel toward the interior. A heavy forklift truck rolling across an upturned corner or edge generates additional stresses, which, when added to the curling stresses, can result in a combined stress sufficiently high to crack a panel.

Floor slabs on ground that are constructed with good materials, designed for sufficient thickness, properly jointed for crack control, and effectively cured will develop strength and other properties sufficient to minimize the tendency to curl. Heavy amounts of reinforcing steel placed 50 mm (2 in.) down from the surface reduce edge curling, too.

Curling may diminish with age as moisture content and temperatures equalize throughout the slab. In addition, creep (relaxation of the concrete) probably reduces curling over a period of months. Curling can be corrected by grinding the slab after filling voids beneath the upward-curled sections. See Repairs for Loss of Slab Support later in this chapter.

Popouts

A popout is a conical fragment that breaks out of a concrete surface, leaving a hole. The hole varies in size from 5 mm to 50 mm (1/4 in. to 2 in.), though larger popouts are possible (Fig. 10-6). Usually, a fractured aggregate particle is located at the bottom of the hole. The matching piece of the fractured particle adheres to the point of the popout cone. Popouts are considered a cosmetic detraction and generally do not affect the service life of the concrete.

For interior floors, popouts may occur to relieve pressure created by alkali-silica gel. Gel forms during the chemical reaction between the alkali hydroxides in the concrete and reactive siliceous aggregates, then expands as it takes up water. Popouts caused by alkali-silica reactivity (ASR) may occur as early as a few hours to a few weeks or even a year after the concrete is placed.

For exterior slabs, popouts are usually caused by a piece of porous rock having a high absorption/low specific gravity, such as pyrite, hard-burned dolomite, coal, shale, soft fine-grained limestone, or chert. As the offending aggregate absorbs moisture, its swelling creates internal pressures sufficient to rupture the concrete surface. Freezing under moist conditions can also lead to rupture.

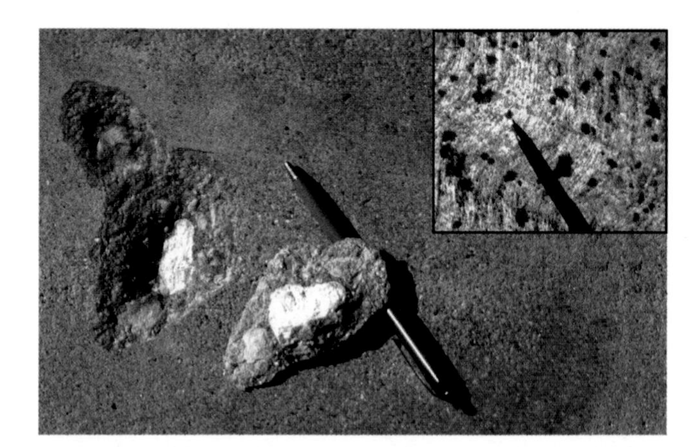

Fig. 10-6. (primary) A popout is a small fragment of concrete surface that breaks away due to internal pressure, leaving a shallow, typically conical, depression. (inset) Popouts caused by sand-sized particles are small and are the most common type found on interior floors covered with sheet tile. (0113, 51117)

Popouts caused by moisture-induced swelling may occur shortly after placement due to the absorption of water from the plastic concrete, or they may not appear until after a period of high humidity—like heavy rainfall—or after the concrete has been exposed to freezing temperatures. Popouts caused by alkali-silica reactivity also occur on exterior slabs.

The following steps can be taken to minimize or eliminate popouts caused by alkali-silica reactivity (Farny and Kosmatka 1997, Landgren and Hadley 2001):

1. During hot, dry, and windy weather, cover the surface with plastic sheets after screeding and bullfloating to reduce evaporation before final finishing. This reduces the migration of alkalies to the surface due to drying, which would concentrate the alkalies and increase the potential for alkali-silica reactivity.
2. Use wet-curing methods such as continuous sprinkling with water, fogging, ponding, or covering with wet burlap soon after final finishing. Wet cure for a minimum of 7 days. Avoid plastic film, curing paper, and especially curing compounds, as they allow an accumulation of alkalies at the surface. Flush curing water from the surface before final drying. Impervious floor coverings or membranes should be avoided.
3. Use blended cement, fly ash, slag, calcined clay, or silica fume that has been proven to control alkali-silica reactivity. Use of a low-alkali cement or lithium admixture is also beneficial.
4. Use concrete with the lowest water content and slump consistent with the application.

Surfaces with popouts can be repaired with small patches. Drill out the spalled particle and fill the void with a dry-pack mortar or other appropriate patch material. If the popouts in a surface are too numerous to patch individually, a thin-bonded concrete overlay may be used to restore serviceability (PCA 1985, PCA 1996a). Proprietary self-leveling toppings can be used for this purpose.

Fig. 10-7. Random cracking. (69672)

Random Cracking

Random cracking, which may occur almost immediately (during joint sawing operations) or at any later time, is more severe than crazing or plastic shrinkage cracking (see Fig. 10-7). If the cracks appear later, it may be a result of normal concrete shrinkage combined with poor jointing practices: not enough joints, poor location, or poor construction. It may also be due to improper subgrade support that allows differential settlement of adjacent sections of the slab.

When random cracking occurs on a newly placed slab, it is usually related to improper timing of joint sawing. The purpose of cutting the slab is to induce a crack beneath the cut. However, it is possible that the slab will crack in front of the cut instead of beneath it, as intended.

Concrete needs to gain adequate strength before having joints cut into it. Ideally, the tensile strength holds the slab together. The sawcut notch creates a reduced slab section, which increases tensile stress in the concrete below the notch. In the reduced section, the tensile stress is of greater magnitude than the concrete tensile strength. Thus a crack occurs below the notch. The crack and sawcut combine to relieve the stresses and thus prevent unwanted random cracking. But new concrete is always trying to shrink. As the sawblade cuts a joint in the concrete, the sawcut weakens the concrete slab. If sawcutting is started when contraction stress (as a result of concrete shrinkage) is great and tensile strength is not yet adequate to resist it, cracks can jump ahead of the blade during joint cutting.

If cooling water (used with wet sawing) hits the warm slab, it can be a thermal shock that adds to the potential for random cracking ahead of the saw blade.

To avert random cracking, sawcut jointing must be done before concrete cooling and drying starts, but after some (tensile) strength has developed. The notch installed by sawcutting should be deep enough so that the crack occurs below the sawcut. Usually, 1/4 of the slab thickness is sufficient. Spacing of joints to minimize random cracking should be about 24 to 30 times the slab thickness. For example, a 200-mm (8-in.) thick slab requires a spacing of between 4800 mm (192 in. = 16 ft) and 6000 mm (240 in. = 20 ft). Whereas shorter spacings will provide better crack control, they also increase the overall number of joints (a maintenance concern). Longer spacings can be used but may not adequately control random cracking within the panel.

Discoloration

Surface discoloration of concrete flatwork can appear as gross color changes in large areas of concrete, spotted or mottled light or dark blotches on the surface, or early light patches of efflorescence.

No single factor is responsible for all discoloration. Some factors that influence whether or not a surface discolors are calcium chloride admixtures, hard-troweled surfaces, inadequate curing, and changes in the concrete mix.

Fig. 10-8. Discoloration. (14019)

Calcium chloride admixture can be used to accelerate cement hydration. The presence of calcium chloride, however, retards hydration of the ferrite compound in portland cement. Since the ferrite phase generally lightens with hydration, if it is retarded, it can remain unhydrated and dark.

Repeatedly steel troweling a floor creates a hard, dense surface that is resistant to wear and abrasion. If, as a result of the steel troweling, the water-cement ratio is drastically decreased in localized areas, the concrete in that location will be darker than surrounding areas.

Waterproof paper and plastic sheets used to moist-cure concrete containing calcium chloride have been known to give a mottled appearance due to the difficulty in placing and keeping a cover in complete contact with the surface. The places that are in contact with curing sheets will be lighter colored than those that are not touching the sheets (Greening and Landgren 1966).

A rare discoloration ranging from buff to red/orange has been reported in Wisconsin, Illinois, Louisiana, and other states. This type of discoloration is more likely to occur during periods of combined high relative humidity and high ambient temperature combined with certain types and amounts of wet curing. Fly ash aggravates the staining by intensifying the color.

This staining is probably caused by differences in curing and degree of hydration of the surface cementitious materials. In particular, additional hydration of the ferrite compounds in cementitious materials leads to more reduced iron being available to oxidize and discolor the concrete. Research has found that the staining occurs under a certain set of conditions, which includes the availability of water and oxygen. Rapid drying of the concrete results in insufficient moisture to produce staining, while continuous immersion does not allow access of air. In both cases, discoloration does not occur. Therefore, it is recommended that the concrete be kept fully wet for the required curing period, then allowed to dry as rapidly as possible thereafter. For instance, wet burlap with a plastic sheet covering should be removed in the morning of a hot day rather than in the evening or before rain is expected.

This type of buff to red/orange staining is difficult to remove. The following chemicals are largely ineffective: hydrochloric acid (2%), hydrogen peroxide (3%), bleach, phosphoric acid (10%), diammonium citrate (0.2 M), and oxalic acid (3%). Commercial sodium bisulfate cleaners are somewhat successful in removing the stain; however, within several weeks after cleaning, the stain reappears (Miller, Powers, and Taylor 1999, and Taylor, Detwiler, and Tang 2000).

Changes in concrete materials or proportions affect concrete color. The sources of sand and cementitious materials should not change if the color of the floor is to remain consistent. And while the source of water will be easy to keep the same, the amount of water mixed with each batch can vary, resulting in potential color changes. With given materials, the concrete having a higher water-cement ratio will generally be lighter in color than the concrete having a lower water-cement ratio.

Efflorescence. Efflorescence is usually a white deposit on the surface of concrete, which can occur just after placing the floor. Efflorescence is a product of the combination of the following three factors:

- soluble salts in the material (carbonates of calcium, potassium, and sodium; sulfates of sodium, potassium, magnesium, calcium, and iron; and bicarbonate of sodium or silicate of sodium)
- moisture to dissolve these salts
- evaporation or hydrostatic pressure to move the moisture toward the surface

Efflorescence is more common in winter when a slow rate of evaporation allows migration of salts to the surface. In warmer weather, evaporation occurs so quickly that comparatively small amounts of salt are brought to the surface.

To avoid efflorescence, sand should be washed and mix water should not contain harmful amounts of acids, organic material, minerals, or salts. Seawater should not be used.

The floor should be cleaned promptly to remove the efflorescence before the calcium hydroxide reacts with carbon dioxide and forms an insoluble calcium carbonate. Acidic cleaners can remove the calcium carbonate but they may etch the surface. If a glossy look is desired from the beginning, an acrylic curing compound or a liquid surface sealer can be used.

By using properly graded aggregate, an adequate cement content, a low water-cement ratio, and thorough curing, the concrete will have low absorption and maximum watertightness, and therefore minimal efflorescence. If a whitish deposit does appear on the surface of the concrete, it can be cleaned with dry brushing, waterblasting, gritblasting, or by scrubbing with a mild acid solution. Treating the entire surface helps avoid uneven surface colors. See Related Literature, *Concrete Slab Surface Defects: Causes, Prevention, Repair* (IS177), for detailed information on efflorescence.

FLOOR MAINTENANCE AND REPAIRS

General maintenance does much to assure long-term performance of any concrete floor. Some repairs may be necessary from time to time. The bulk of the attention is given to protecting the floor surface and the proper functioning of joints. When maintenance and repairs are performed in a timely manner, the floor:

- is able to carry loads it was designed to carry
- protects goods stored on it
- can remain in service without interruptions to traffic flow

For a well-designed floor installed with good workmanship and good materials, maintenance will primarily be focused on minor repairs and cleaning. Repairs include patching spalls at slab edges or joints, routing and sealing or filling of random cracks, and the restoration of areas damaged by chemical attack from spillage or impacts from heavy materials. The preceding maintenance examples are to be differentiated from housekeeping maintenance done to assure cleanliness, safety, or compliance with regulatory agency requirements (if applicable). Regular washing of floors can reduce wear by abrasion (by removing grit).

Unfortunately, not all floors are designed and constructed to ideal standards, nor does the design use necessarily prevail for the life of the floor. When floors are exposed to uses other than originally intended—resulting in increased loads or additional traffic—it is possible that damage will occur and require substantial repairs. Also, some facilities serve beyond their anticipated life, then need rehabilitation to bring them to a serviceable condition. Repair and rehabilitation requirements can be extensive, so only the more common repair methods are discussed here.

Planning the scope of repairs and rehabilitation should be done in consideration of:

- future use of the facility—loading and traffic
- final elevation of the floor surface (proposed repairs can change floor height)
- maintenance and repair budget
- operations and downtime—can facility be shut down during repairs or must it continue to operate?
- scheduling—how long will floor be occupied during repairs?
- protection of adjacent work and/or storage areas

The basic principles of successful floor repairs include determining the cause and extent of damage, careful preparation of the old concrete, the placement and curing of high-quality replacement concrete, and proper joint treatment. The following information and recommendations should be used as a guide to obtaining satisfactory results.

Evaluation of Deficiency

Evaluation of deficiency is necessary to determine the cause or causes for observed distress; from this information, a repair method can be determined. This helps ensure that the repair will be effective and the deficiency will not extend into the concrete surrounding the repair area. The following list is not intended to be all-inclusive, but should provide some of the factors that need to be evaluated in order to design and plan a repair:

- slab thickness
- slab loading
- random slab cracks
  - crack width
  - spalling at cracks
  - location of reinforcement
  - warping at cracks
- spalling at joints
  - depth of spalls
  - width of spalls
  - width of crack at joint
  - load transfer dowels
- active and non-active joints
- joint sealants
- warping at joint
- warping at corners
- load transfer dowels
- loss of slab support
- surface hardness-dusting
- surface scaling
- surface blistering
- surface delamination
- popouts
- surface gouging
- surface levelness
- surface flatness

For best results, an engineer with expertise in floor design, construction, performance, and repair should be hired to (1) do an evaluation, (2) design and specify repairs, and (3) supervise repair work.

Evaluation Methods

Prior to repair, an inventory should be made of floor deficiencies. This helps define the scope of work and strategies for accomplishing repairs. The listing below serves as a starting point for assessing the condition of the floor. Not every test is needed for every floor. On the other hand, additional tests may be needed for some floors. The extent of the evaluation depends on the experience and judgment of the engineer in charge of the repair. Activities for the evaluation and the results they provide may include:

- visual examination
- surface elevation profile measurements to quantify flatness, levelness, and upward warping deformations at joints (ASTM E 1155)
- sounding for surface delaminations or debonding of toppings (ASTM D 4580)
- coring slab for strength testing and to confirm results from sounding (ASTM C 42)
- coring at cracks and joints
- petrographic examination (ASTM C 856) to determine
  - concrete quality
  - near-surface characteristics
  - reactive aggregates
  - depth of contaminants
  - depth of removal (bonded repair toppings)

- ground penetrating radar survey to determine
  - location of steel reinforcement in floor slab
  - alignment (or misalignment) of load transfer dowels
  - slab thickness (especially where deficient)
  - presence of voids beneath the slab
- pachometer measurements to determine the depth and location of reinforcement in the slab
- plate bearing tests (on top of subbase/subgrade) to determine modulus of subgrade reaction (ASTM D 1196)
- slab thickness (ASTM C 174)
- cone penetrometer measurements (made through core holes) to determine subgrade compaction density (ASTM D 4633)
- pulse velocity to indicate the presence of voids or cracks (ASTM C 597)

Repair of Cracks and Joints

Although joints are placed in concrete floors to provide crack control—that is, to encourage the location of slab cracks (from shrinkage or other contraction stresses) beneath the sawcut alignment—random slab cracks, as shown in Fig. 10-7, occur in some floors. Random cracks are usually the result of the following causes, alone or in combination with each other:

- distance between contraction joints (sawcuts) is too great
- sawcuts are made too late, that is, *after* restraint stresses have exceeded concrete tensile strength
- isolation joint is not properly constructed (for example, reinforcement is carried from walls into floor)
- slab reinforcement is continued through joints
- joint notch is of insufficient depth (either sawcut or formed joint)
- dowel alignment is poor, causing restraints to slab end movements
- loads exceed as-constructed floor load capacity
  - inadequate subgrade support and/or soil consolidation or settlement
  - design error
  - inadequate concrete flexural strength
  - overloading during construction

Joints and/or cracks must be prepared before they can be repaired.

Tightly closed cracks and fine cracks subject only to light industrial traffic should be left alone, but kept under observation. They usually do not affect floor serviceability.

As concrete contracts (shrinks), cracks can open. If cracks open up or show signs of spalling, they should be filled (see Fig. 10-9). Spalling occurs at widened (opened) cracks subjected to hard rubber or solid tires, hard polyurethane casters, or steel wheels. The unsupported slab edges at the crack cannot support the wheel loads, so they break off, or spall. If not repaired when first observed, the spall becomes wider and deeper because wheel traffic loads gain an impact component. Thus, what is initially a relatively simple repair of routing and filling can develop into a more complex and costly partial depth repair, as shown in Fig. 10-10, or full depth joint repair, as shown in Fig. 10-11.

In concrete slabs, expansion and contraction occur regularly as a result of temperature and moisture fluctuations. These ongoing slab movements may have to be accommodated after the cracks have been repaired. If nearby joints are inactive (do not accommodate slab end movements), provisions need to be made as part of the repair to continue to accommodate the movements at the crack. Otherwise, the crack repair will fail when the slab movements occur again.

Semi-rigid joint fillers have very limited expansion capability and may separate adhesively or cohesively when movement occurs (see Fig. 10-12a and b). Repair may be needed if the filler becomes loose or ineffective in preventing spalls.

Elastomeric sealants can also fail when subjected to joint movement (see Fig. 10-12c). Failed sealants cannot be repaired and require removal and replacement.

Spalling at joints, like at cracks, is usually attributed to traffic across the joint. Spalling occurs because the edge of the slab (a vertical face) is not supported laterally. The

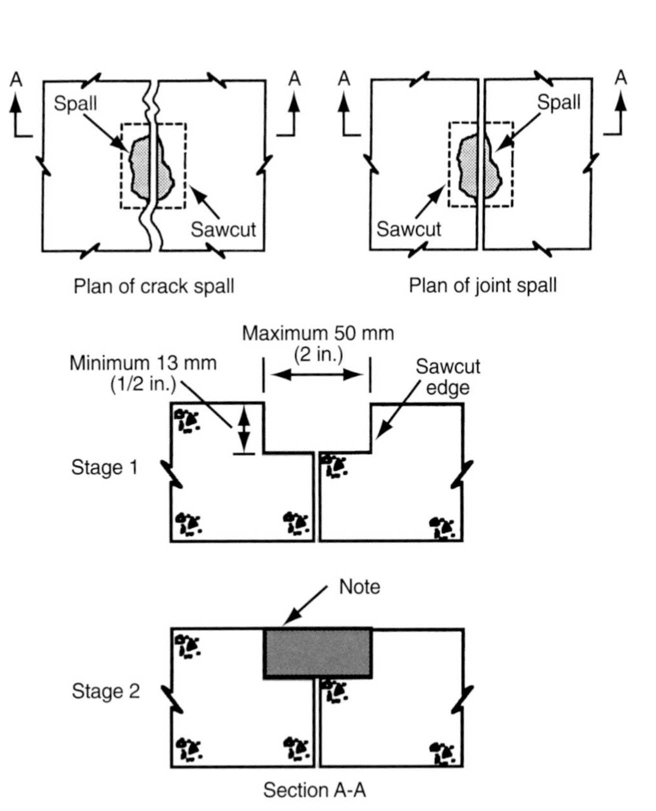

Fig. 10-9. Joint and crack routing repairs. Section A-A. Note: Spalls in floors subject to hard wheel traffic should be repaired with semi-rigid filler as shown. Some filler manufacturers recommend adding silica sand to filler for spalls greater than 19 mm (3/4 in.). Follow manufacturer's recommendations.

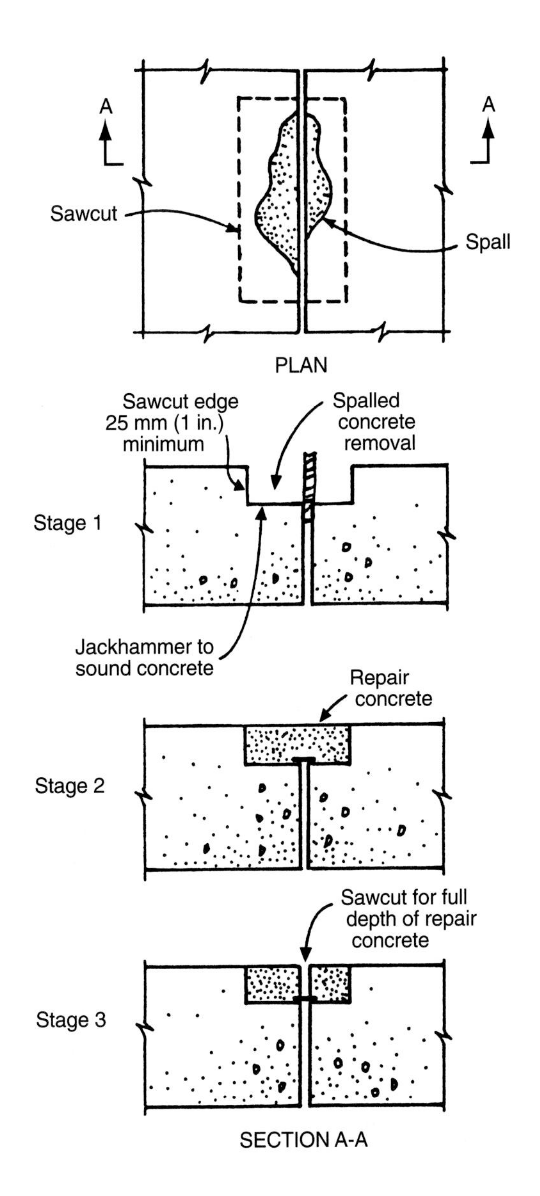

Fig. 10-10a,b. Partial depth repair. (69676)

immediate objective of spall repair at a joint is to restore the distressed slab edge, and the long-term objective is to avert future performance problems.

In effect, a crack is a self-determined joint in the concrete slab. Because cracks and joints are very similar, repairs for cracks and joints share many similarities and are discussed together in the following material. In the next few sections regarding repair, the terms crack and joint are used interchangeably unless noted otherwise.

Inactive crack/joint repair. For crack and joint repairs where there is no movement, the spalled concrete is removed by routing out the crack or joint with a saw and gluing the two sections back together. Preparing concrete by routing includes removing all unsound concrete, dust, and debris. A structural epoxy can be used to bond the two crack faces, and an epoxy concrete repair mortar can be installed in the routed-out space. If stipulated by the epoxy mortar supplier, a primer is applied to the concrete repair surface.

Active crack/joint repair. If slab movements are to be accommodated at the joint or crack, the repair is done with a semi-rigid epoxy. This will provide support to the vertical edges, but still allow expansion or contraction movements without damaging the repaired area. Routing in preparation for installing the semi-rigid epoxy should be to a depth of about 13 mm to 25 mm (1/2 in. to 1 in.) and as narrow as practical.

Partial- and full-depth repairs. Deeper and wider spalls require deeper and wider repairs. In every case, the repair must extend to sound concrete. If depth to sound concrete is greater than one-half of the slab thickness, full-depth repairs must be made.

For partial-depth repairs, the outline of the repair area is marked. A sawcut to a minimum 50-mm (2-in.) depth is made around the edge of the repair area, and concrete is removed by chipping or scabbling. After reaching sound concrete, the repair area is then blasted with grit or shot. Ganged diamond saw blades or cylindrical drums impregnated with diamonds can be used instead of the impact removal methods. After surface cleaning, a neat cement grout or epoxy bonding agent is applied to the repair surface. A concrete repair mix or an epoxy concrete is then placed over the bonding layer before it dries. If movements are to be accommodated at the joint, the joint must be carried to the slab surface for the full repair depth. A form installed over the joint can serve the purpose. This should be removed after hardening of the repair material, and the space filled with a semi-rigid epoxy.

For full-depth repairs, the slab is sawcut to full depth to include all spalled and unsound concrete in the joint vicinity. The minimum width and length of repair area is about one meter (one yard) in both directions to accommodate dowels. After making full-depth sawcuts, horizontal (dowel) holes are drilled into the vertical sawcut face at all four sides of the opening. The subbase in the repair area should be recompacted and the surface graded to allow the proper slab thickness. Repairs that cross an active joint or crack must not interfere with the joint function. Fig. 10-11 (Section A-A) shows that repairs are made first to one side of the joint, then to the other. Bond is prevented between the two joint faces. A fillerreservoir (aligned with the joint) must be formed or cut and sealed with a semi-rigid filler. Dimensions of the reservoir are about 6 mm (1/4-in.) wide by 40 mm (1-1/2 in.) deep.

The elevation of the repaired concrete surface must match that of the surrounding floor. A concrete mix should be installed, moist-cured, and protected from traffic until the repair material reaches the specified strength of the original concrete (floor).

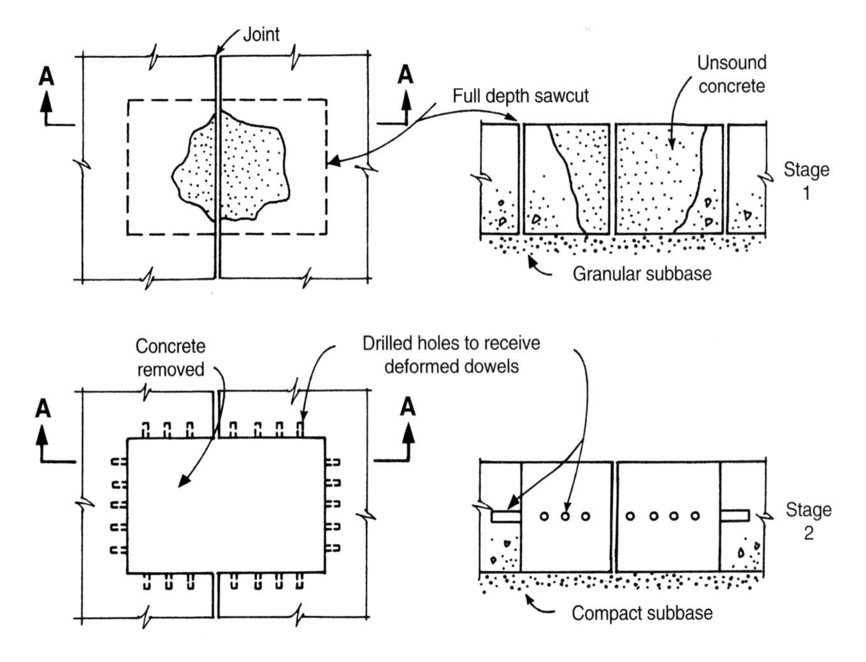

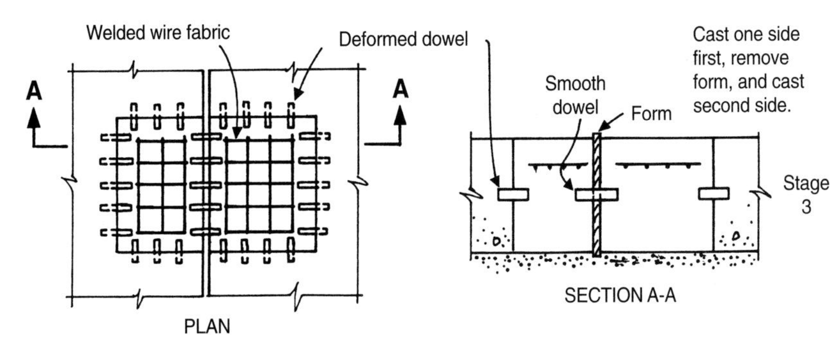

Fig. 10-11. Full depth joint repair.

Repairs for Loss of Slab Support

Loss of slab support can be attributed to:

- upward warping deformations at slab edges and corners (curling)
- densification of subgrade and subbase due to slab vibration under traffic
- pumping (displacement of relatively soft/moist cohesive soils due to slab edge deflections under repeated traffic)
- soil shrinkage and settlement due to reduced moisture content (caused by droughts or reduced water tables)

Large edge deflections can be attributed to:

- insufficient slab thickness
- inadequate load transfer efficiency at the joint

As a result of loss of support or repeated large edge deflections, a void or soft area occurs below the slab edge. If left unattended, a crack parallel to the joint can occur.

Pumping grout under the slab to fill the void, often called "undersealing," will restore slab support and can reduce large deflections. In some cases, soft cohesive soils will again yield under repeated deflections. Then the potential cracking problem (parallel to the joint) is not resolved, only postponed. For certain areas, establishing load transfer across joints is the only way to reduce deflections.

Dowel Retrofit. If the loss of support occurs in isolated locations, load transfer can be established by filling the joint and crack below the sawcut with a rigid epoxy. If, however, the problem occurs at successive joints for an appreciable length of the floor, load transfer is established by installation of dowels–smooth steel bars. To install dowels for a joint repair, these four steps are followed:

1. slabs are cut transverse to the joint and the concrete is removed
2. holes are drilled into concrete faces at mid-depth
3. dowels are placed in the joint (aligned perpendicular to the joint face)
4. the slot is filled with an epoxy concrete (interior slabs only)

Instead of individual dowels, a proprietary load transfer device can be used to establish load transfer. Whether individual dowels or proprietary load transfer devices are used, (at least) one end of the smooth steel dowel surface should be coated with a bond breaker. When the patch material is placed, the dowels should be completely surrounded by new concrete and have good bearing.

Undersealing. If large areas of the slab are affected and require stabilization, a grout mix consisting of cement and fly ash is pumped into the void below the slab. Using opti-

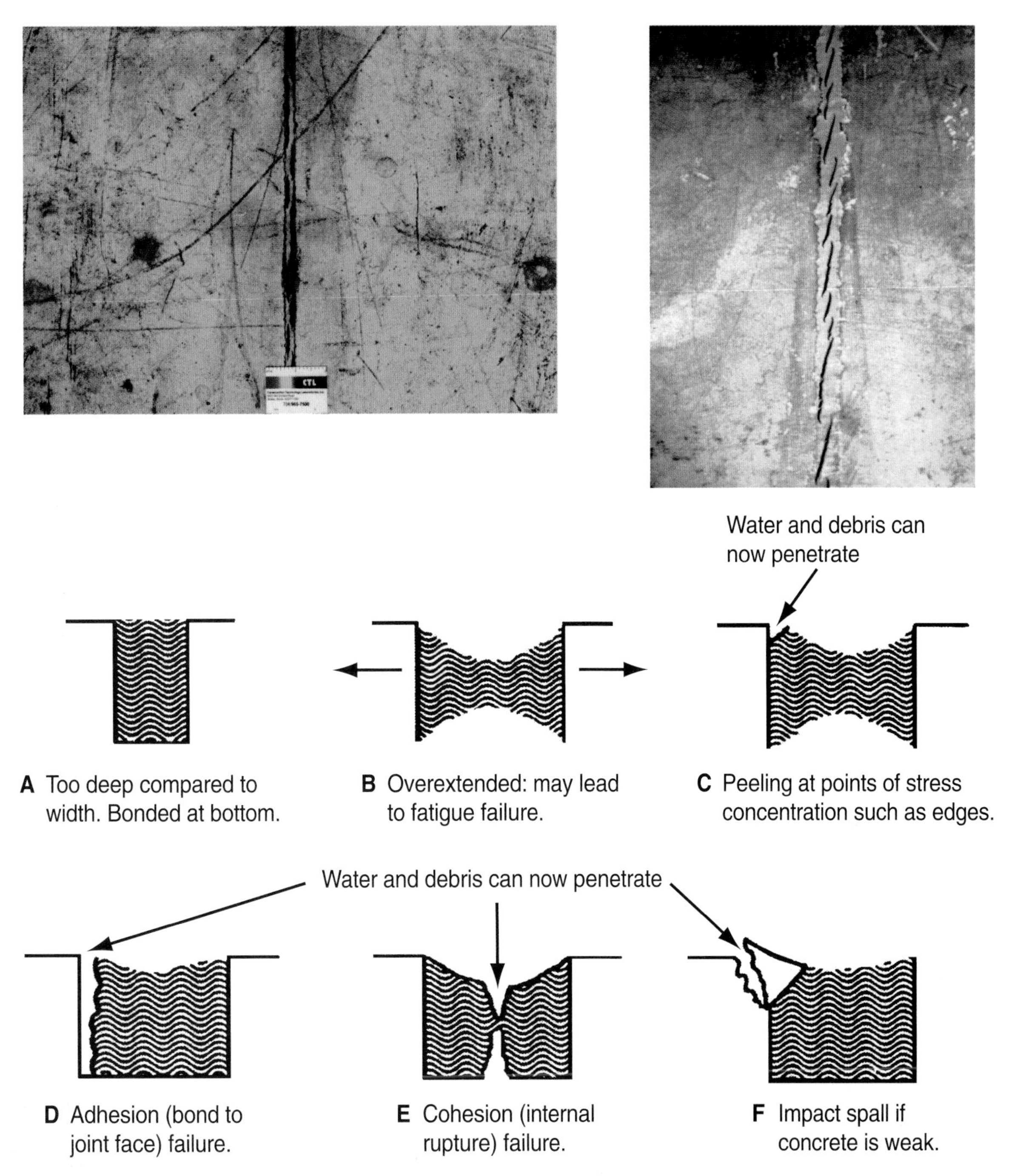

Fig. 10-12. (top left, top right) These joint fillers did not accommodate all of the joint movement. Semi-rigid fillers are not expected to do so, but correction may be required if the filler ceases to be effective in preventing joint edge spalling. (67201, 69680). (bottom) Joint sealants must be installed to the proper thickness and depth. Sealants must be able to stick to concrete, stretch, and compress to accommodate joint movements, but semi-rigid fillers used in interior slab joints are not expected to. (ACI 504 R)

cal or mechanical instrumentation, the elevation (grade) of the slab surface should be monitored during the pumping operation. Care must be exercised not to raise the slab.

Holes through the slab for introducing the grout mixture are spaced at about 0.6 m (2 ft) on centers and about 0.3 m to 0.6 m (1 ft to 2 ft) inward from the joint—depending on dimensions of the void below the slab. Low-pressure pumping is used to stabilize slabs. Since the goal is not to lift the slab but merely stabilize what is under it, required pressures for grouting are low—from gravity feed to around 0.15 MPa (25 psi)—though higher pressures may be used. In any case, the grout pump should be equipped with a pressure cutoff control as excessive pressure surges can cause slab lifting. Grouting is continued until grout flow is observed at adjacent holes and at the joint. Upon completion of grouting, the joint space should be cleaned to remove all grout and laitance in preparation for joint sealing.

Grinding. Power grinding is done both on new and existing slabs. Grinders can be used to remove weak, friable laitance; improve levels; and remove high spots and trowel and other marks. Grinding also can be used to produce a smoother floor. The wear resistance of the surface may be increased or decreased by grinding, depending on the finishing techniques and aggregates.

If a surface is soft or abrasion-prone, it may wear and produce dust under traffic. Grinding is a technique used to improve this kind of surface by removing a thin layer of concrete, about 1 mm to 1.5 mm (1/32 in. to 1/16 in.) or more. For the grinding to be effective, the underlying concrete must have sufficient resistance to wear.

Grinding can also be used to correct slab curling. Curling, or warping, occasionally occurs on slabs, usually near joints or other slab edges.

Where upward warping deformations of slab edges interfere with operating conditions of the facility, surface grinding can restore the desired flatness. In most instances, it is only necessary to grind inward from the joint a distance of about 0.6 m to 1.0 m (2 ft to 3 ft). Grinding can be done with a terrazzo grinder equipped with a diamond-impregnated disk, or if the magnitude of warping is severe, by localized rotomilling followed by grinding. Rotomilling should not be done if it causes joint spalling. It should be recognized that grinding will alter the appearance and properties of the concrete surface. Coarse aggregate will be exposed at the surface, similar to a terrazzo surface. In addition, the surface hardness of the ground area may be less than that of the adjacent hard troweled surface, depending on the finishing techniques and aggregates.

Concrete Floor Surface Repair

Concrete surface repairs can range from small areas with a localized deficiency—extending over less than a few panels—to restoration of the entire slab-on-grade floor within a facility. In most instances, repairs must be made without changing the original slab surface elevation. Operation of lift trucks generally makes it undesirable to use transition slopes and ramps to accommodate overlays of varying thicknesses (as occurs when the depth of scarification of the existing floor is not uniform). For most repairs and restorations, the choice of floor surface replacement falls between complete removal and replacement of the floor to installing a bonded topping. If the installation will only be exposed to foot traffic and can tolerate slight slopes to adjust surface elevations, an unbonded overlay can be used.

Following are some of the reasons to consider floor surface repairs or restoration:

- dusting and rapid surface wear
- a surface deficiency (delamination, blistering, other)
- abrasive wear in wheel paths
- surface gouging, as occurs from heavy abrasion (from equipment, pallets, or skids)
- surface erosion (chemical attack, moving water)
- to replace wood block or tile surfaces with concrete surfaces
- to overlay asphalt floors

Surface preparation is vital to production of a satisfactory concrete repair. Irrespective of materials, method, system, or thickness to be used, it is essential to achieve a clean, strong bonding surface with a suitable mechanical key that is free of contamination, laitance, and dust. There are many suitable methods of surface preparation, some of which are described in the following section.

Concrete Floor Surface Preparation

In preparing a concrete floor to receive topping repairs, one or more methods may be used. These include wet and dry methods. The best equipment/method for the job will depend on a number of factors:

- thickness of material to be removed
- condition of the concrete (sound, cracked, soft)
- presence of steel reinforcement close to the surface
- type and thickness of new coating or topping
- allowable noise, dust, exhaust fumes

Selection of method also depends on the type and severity of contamination, and the magnitude of bond pull-off strength needed for successful performance of the bonded topping. Testing for adequacy of surface preparation is discussed under bond pull-off testing on pages 88 and 89.

Wet cleaning. By wet cleaning, it may be possible to bring a floor to an adequate surface condition to accept a bonded topping. Two materials, one acid-based and the other alkali-based, are commonly used in three separate operations: (1) degreasing, (2) acid treatment, and (3) neutralizing. Cleaning operations should be performed only under conditions where appropriate safety precautions are taken.

Degreasing is accomplished by mixing a cleaner and curing-compound remover (a chlorinated, emulsifiable solvent), an industrial grease remover (a highly alkaline, low-phosphate, biodegradable detergent), and liberal amounts of water. The mixture is used to scrub the surface—repeatedly, if necessary—for added cleaning effect. The surface finally is rinsed and scrubbed with clear water, vacuumed damp-dry, and then allowed to air-dry.

Acid treatment of the surface is accomplished by applying an etching solution (a mild organic acid combined with detergents, emulsifiers, and solvent) to the *dry* floor. The solution is scrubbed over the floor, followed by a vacuum machine to pick up residual material. The surface is then rinsed, scrubbed clean of acid solution, and dried.

Neutralization of the surface for any residual acid is necessary. This is done by wetting the surface with clear water, sprinkling on a detergent cleaner (highly alkaline, high-phosphate, non-residue detergent), scrubbing, rinsing with water, and vacuuming damp-dry. The surface is then rinsed again with clear water, vacuumed damp-dry, and allowed to air-dry before applying the new surface finish.

An acid treatment is recommended only for cleaning floor surfaces that are to receive a new surface finish. Acid treatment of floors has become less and less common because the acid cleaners require strict handling and disposal procedures for safety.

Other wet cleaning methods include high-pressure water-jet blasting and steam cleaning (see Fig. 10-13) (PCA 1995a). Tough cleaning jobs require the use of heat (up to about 95°C [200°F]) and chemicals. Water blasting can be done with or without abrasives and at varying water pressures. In addition to water, abrasives may be used to improve the cutting action. Water blasters can be used to scarify the concrete surface, removing only 6 mm (1/4 in.). At the other extreme, water jets can remove concrete depths up to 300 mm (12 in.). Operating pressures vary from 0.6 MPa to 69 MPa (100 psi up to 10,000 psi), with 14 MPa to 21 MPa (2000 psi or 3000 psi) being common for cleaning. Steam cleaning can be done with or without chemicals. But because it is slow, requires high water temperatures, and does not do as good a job as pressure washers, steam cleaning slabs is generally not economical.

Dry preparation. Mechanical methods offer an efficient and economical way to dry clean a concrete surface. Equipment is available for scabbling, scarifying, grinding, shotblasting, planing, and flame cleaning. Some machines have a variety of interchangeable cutting and brushing heads.

Most scabbling and scarifying machines are mechanized versions of a hammer and chisel. Scabbling uses an air-driven machine that has a varying number of pistons mounted in a block. Each piston is fitted with a tungsten carbide bit. Operating much the same as a bushhammer or chipping hammer (Fig. 10-14), the striking action shatters away the surface contamination, exposing a clean, sound concrete with a good mechanical key for bonding.

Scarifying uses a power-driven tool with rotary heads that can be fitted with a wide variety of scrubbing brushes, cutting wheels, scouring brushes, and steel scarifying brushes. Large open areas are suited for scarification by machine. Scarifying and scabbling tend to roughen the surface more than shotblasting or abrasive blasting.

Shotblasting is done by closed-cycle steel-particle bombardment to remove any surface contaminants. Shotblasting is virtually dust free and is frequently used to prepare floors for installation of coatings.

Planing machines can remove laitance, paint marks, pitch adhesives, and thermoplastic adhesives, and can cut back the level of the concrete. Other uses include grooving the concrete and making nonskid surfaces. The machines have hardened-steel cutting wheels for hammering off the surface.

Scabbling, bushhammering, rotomilling, and chipping cause impacts at the concrete surface (see Fig. 10-15). These impacts can cause horizontal or near-horizontal cleavage planes that reduce concrete tensile strength at the surface. To determine if the tensile strength has been reduced, a bond pull-off test is made. Where tensile strength is lower due to the impact scarification process, abrasive blasting or water blasting is required before installing the bonded topping.

Concrete Overlays

When surface wear has reached a point considered detrimental to building operations, action is required to rebuild

Fig. 10-13. Water blasting or jet blasting is an effective method of removing concrete without harming the concrete substrate that remains. (69700)

Fig. 10-14. Weak, defective concrete can be removed with a chipping hammer or jack hammer. (69677)

Fig. 10-15. Shotblasting removes concrete by impelling steel shot towards the surface. (52255)

and restore the original surface. The choice of repair method depends on cost, volume of the repair area, acceptable downtime, required strength, required chemical and abrasion resistance, and ambient conditions while repair is underway.

A bonded topping (also called thin-bonded resurfacing) can be installed using normal or high-strength portland cement concrete mixes or epoxy concrete formulations. With concrete overlays, bonding grouts may be used, but are not always necessary. Concrete overlay mixes that have a high paste content or that are fluid usually do not require a bonding grout. A test patch should be placed to verify the adequacy of bond. When bonding grouts are used, they can either be cementitious (neat cement) or epoxy primers and/or epoxy grouts. Many epoxy materials are not resistant to moisture, so they should not be used where moisture barriers were not installed below the concrete slab (as discussed in Chapter 3), nor anywhere else there is a potential for water vapor pressure in the slab. Some epoxies can be used in moist conditions without concern.

When relatively small areas on the concrete surface require repair, bonded patching is used. The basic principles are the same as for toppings. Both patching and thin-bonded resurfacing are discussed below.

BONDED PATCHING

Surface preparation is the key to producing a well-bonded, shallow concrete patch. Usually, if the patch fails to bond tightly, it is because laitance, dust, or other contaminants were not properly removed from the floor surface.

All weak and defective concrete must be removed, preferably by saw-cutting 20 mm (3/4 in.) deep around the perimeter of the patch, followed by removal of defective concrete with a chipping hammer, a bushhammer, or a scabbler. A chipping hammer should not be heavier than the nominal 7-kg (15-lb) class. Hand tools such as hammers and chisels should be used to remove all final particles of unsound concrete. If needed, gritblasting or water-jet blasting should follow. This step can clean the exposed reinforcement of attached mortar and visible rust. It also cleans all concrete surfaces against which new patch material will be placed. Prior to placing new concrete, the surface should be blown clean by airblast, free of compressor oils, and then flushed with water. Puddles of free water must be removed before placing the patch.

If using a bonding grout (optional), it should be placed immediately prior to placing the new concrete. The cleaned surface should be brushed with a thin (3-mm [1/8-in.]) coat of freshly mixed grout consisting of sand, cement, and water (see Fig. 10-16). This bonding grout should have one part fine sand (2.36 mm [No. 8] maximum size) and one part portland cement mixed with one-half part water to give a creamy consistency. Alternately, the bonding grout could consist of a neat cement mixture (no sand). Whichever grout is chosen, it should be applied at a rate that will keep it from drying or skinning over before it is covered with new concrete.

Mix proportions for patching concrete. When normal or high-early-strength portland cement is used for the repair, the mix proportions by volume for concrete and mortar should be as follows:

Mortar (for patches less than 25 mm [1 in.] thick):
- 1 part portland cement
- 2-1/2 to 3 parts sand

Concrete (for deeper patches):
- 1 part portland cement
- 2-1/2 parts sand
- 2-1/2 parts coarse aggregate

Fig. 10-16. A bonding grout is scrubbed into the prepared concrete surface with a stiff broom. (69678)

For concrete patches, the maximum size of coarse aggregate is limited to roughly one-third the patch depth. For 20 mm (3/4 in.) thick topping patches, the maximum size of coarse aggregate should be 10 mm (3/8 in.). For 50 mm (2 in.) thick patches, the maximum size of coarse aggregate should be 19 mm (3/4 in.).

The materials are proportioned to make a low-slump mix using a water-cement ratio not to exceed 0.44 by weight. Concrete subjected to freeze-thaw cycles and deicer chemicals needs an adequate system of air entrainment with about 5% to 7% total air content (9% for mortar). Where rapid reuse of the patched area is needed, calcium chloride, not to exceed 2% by weight of cement, can be used to accelerate hardening and early-strength gain of non-reinforced concrete. Topping patches must not be feather edged.

Caution: Calcium chloride should not be used when dissimilar metals or electrical conduit are encased in the concrete.

Placing and finishing the concrete. Some type of surface vibration is mandatory for compacting surface patches. A hand-operated vibratory screed or strikeoff is effective, as is a portable plate vibrator or compactor float. Hand tampers can be used when power equipment is not available.

New concrete should be placed and struck off slightly above the final grade and then mechanically vibrated, screeded, and floated to final grade (see Fig. 10-17). When a stiff, low-slump mix is used, there may be no water gain on the surface. Troweling should not be done when there is free water on the surface and should not be done to the extent that free water will be brought to the surface. When the surface has become quite hard, it should be given a second and third troweling to produce a very hard, dense finish.

Curing. Evaporation of water from thin-concrete patches is rapid, and if not prevented immediately after the concrete is finished, rapid evaporation may cause surface crazing, cracking, and curling. Covering the patch with wet burlap prevents rapid drying. The surface should be kept continuously wet for 24 hours, and then covered with polyethylene film or waterproof paper for an additional 48 hours. Ambient conditions (hot or cold, dry or humid, calm or windy) dictate the choice of a satisfactory curing method.

Fig. 10-17. A stiff mix is placed onto a slab area coated with a still-wet bonding grout. (69679)

Thin Bonded Topping

A fully bonded concrete topping is usually at least 20 mm (3/4 in.) thick; an unbonded topping is usually a minimum of 100 mm (4 in.) thick. The thickness of an overlay must be considered because it raises the level of the floor.

The best surface preparation for a bonded overlay is obtained with one of the dry-cleaning methods described previously. If contamination is present after dry cleaning, then wet-cleaning methods should be used to further prepare the surface. An effective check for contamination is to sprinkle water on the concrete: if the water forms little globules, contaminants are present that will interfere with bond. If the water is immediately absorbed, it can be assumed that the concrete is clean. For surfaces contaminated to significant depth by oils—as for example machine building and automotive production plants—petrographic examination will provide guidance on the depth of scarification needed to reach sound concrete that will provide good bond pull-off strength.

Mix preparation for bonded overlays. Volume proportions are suitable for small patches, but for large resurfacing jobs, weight batching should be used to eliminate variations in quality. The mix design varies, depending upon the thickness of concrete to be placed. The water-cement ratio can range from 0.33 to 0.45. Aggregates can be selected for their hardwearing and abrasion-resisting qualities. Low-slump concrete generally works well (50 mm to 75 mm [2 in. to 3 in.]). Adjustments in consistency depend on thickness of topping, temperature, and equipment available to do the work. A water-reducing admixture can improve workability of low-slump concrete, and a superplasticizer can produce high-strength concrete at flowable consistencies.

Formwork. The same principles that apply to edge forms in general apply to topping forms. The method of installing and anchoring varies greatly depending on the job conditions. The forms and intermediate screeds must be set to grade and positively anchored in position to maintain a true level.

Laying the topping. The topping should be placed in panels with panel edges aligned with base slab joints. The topping must be well compacted to ensure bond, strength, and durability. On large projects, a self-propelled finishing machine with vibrating screed supported on guide rails should be used. On smaller jobs, a manually operated vibratory screed or tamper can give satisfactory results.

Various methods of finishing floor slabs have already been discussed. The new surface must be moist cured for at least 3 days (preferably 7 days). One practical and satisfactory method is to cover the surface with polyethylene sheets, with laps sealed and all edges held down. After removing the sheets, an application of curing compound will extend the curing period.

All joints in the original floor must be duplicated exactly in the new topping; that is, they must be located directly over the joints in the original slab and must be of equal or slightly greater width. Duplicating joints can be done without difficulty by placing inserts in the new concrete after screeding the surface. Saw cuts must be timed to avoid reflective cracking from the joint in the old slab below.

Epoxy Concrete Toppings

Epoxy concrete toppings with high wear resistance and high strength can be installed as mortars (2-part epoxy mix added to and intermixed with sand-size aggregate) to 6 mm (1/4 in.) or greater thickness. Epoxy concrete thickness should be 13 mm (1/2 in.) or greater. Surface preparation is as described above for bonded patches and bonded topping. The epoxy systems gain strength rapidly and can be quickly opened to traffic. Manufacturer's recommendations should be followed. The epoxy systems should not be installed on floors with water vapor pressure and water transmission potential.

Unbonded Toppings

The minimum thickness of an unbonded topping is 100 mm (4 in.) for foot traffic; heavier loads require thicker floors. Thickness is determined on the basis of thickness design as discussed in Chapter 5, although a minimum of 150 mm (6 in.) is suggested for floors carrying heavy moving loads and subjected to severe impact. The amount of work required to prepare the old slab is negligible, consisting only of sweeping clean and filling in badly worn areas and holes with a cement and sand mortar.

A separation layer should be used to ensure that no bond occurs between the old and new concrete. A 0.10-mm (4-mil) polyethylene sheet is adequate for this purpose. The separation layer prevents cracks in the old floor from carrying through into the new topping—this is called reflective cracking. It also reduces friction at the interface so that the new topping can move (from drying shrinkage and thermal changes) independently of the floor below.

The concrete mix proportions should be the same as those suggested for bonded toppings or for new floor construction.

Placing and finishing operations are similar to those used to construct a new concrete floor on plastic sheeting. An unbonded overlay permits contraction and construction joints in the old slab to be ignored. The new topping can have a jointing arrangement designed for the most convenient panel size and shape, taking into account the thickness of the topping and the amount of reinforcement provided (if any). However, isolation joints in the old slab must be repeated in the new topping.

Fig. 10-18. Self-leveling floor toppings and underlayments are used to provide a smooth, level floor surface as well as repair floors that have deteriorated, sagged, scaled, or become worn. (69682).

SELF-LEVELING TOPPINGS AND UNDERLAYMENTS FOR FLOORS

Many proprietary thin toppings and underlayments are available for floor applications. They are often referred to as self-leveling (nontroweling) or trowelable materials (see Fig. 10-18). These materials are primarily used to provide a smooth, level floor surface, as well as repair floors that have deteriorated, sagged, scaled, or become worn. These materials may reduce sound and heat transmission between slabs as well as provide additional fire resistance.

Toppings provide the actual wearing surface of a floor and have been previously discussed under the headings of thin bonded concrete and epoxy concrete toppings. These are appropriately used when the topping surface carries traffic or when the topping is selected for its chemical resistivity. Certain topping materials can be used outdoors.

Underlayments can be provided when a material such as tile or carpet is used as a floor covering. Particular underlayments are formulated to be applied over old or new floors constructed of cast-in-place concrete, precast concrete, wood, tile, and terrazzo, as well as floors contaminated with adhesive residues. Some can be placed over radiant-heating pipes. Most underlayments are for interior use.

Composition

Underlayments are composed of a combination of cementitious materials, sand, air, water, and admixtures. Sometimes coarse aggregate, polymer modifiers, or fibers are used. The dry materials are usually blended together and packaged in bags for easy use on small projects.

Lightweight cellular concrete is an underlayment material (floor fill) containing normal grout or concrete ingredients (minus the coarse aggregate) along with a foaming agent that produces a high air content. The high air content provides low weight and added fire and sound resistance. It is usually centrally batched using normal bulk materials. More information on lightweight concretes for these and other applications can be found in *Design and Control of Concrete Mixtures* (EB001), available from the Portland Cement Association; see Related Publications. For further information on low-density concrete, see References ACI 523 (1992) and ACI 523 (1993).

Properties

Self-leveling underlayments are formulated for minimal shrinkage, rapid strength gain, and flowability without segregation. However, these and other properties can be adversely affected by an excessive amount of water. Too much water will cause dusting, low strength, and shrinkage cracks. Self-leveling toppings and underlayments must be pumpable and be able to level off after minor screeding. Trowelable mixes have a stiffer, less-watery consistency.

These materials weigh from 1600 kg/m^3 to 1920 kg/m^3 (100 pcf to 120 pcf) dry and have compressive strengths of 7 MPa to 52 MPa (1000 psi to 7500 psi) after 28 days. Cellular concrete has a minimum strength of 7 MPa (1000 psi) for residential applications or 10 MPa (1500 psi) or more for commercial underlayments. The strength of cellular concrete can be increased to about 21 MPa (3000 psi) or more by reducing air and sand content. Many noncellular underlayments range in strength from 20 MPa to 40 MPa (3000 psi to 6000 psi).

The shrinkage characteristics of proprietary underlayment products can vary considerably; therefore the manufacturer should be consulted about joint spacing and crack control. Joint spacings commonly range from 3 m to 9 m (10 ft to 30 ft). Joints in the base slab must be duplicated in the underlayment.

Underlayment thickness, minimum and maximum, depends upon the individual product and where or how it is used. The range of allowable thicknesses varies from 3 mm to 100 mm (1/8 in. to 4 in.). Thin applications—less than 25 mm (1 in.) deep—are usually bonded. Thicknesses greater than 13 mm to 38 mm (1/2 in. to 1-1/2 in.) have aggregate added to reduce shrinkage. Minimum thickness over wood will be more than over concrete. The minimum thickness for a cellular concrete fill is about 40 mm (1-1/2 in.) or 20 mm (3/4 in.) if bonded over precast concrete. Cellular concrete is placed by bonding directly to the concrete base slab or is installed unbonded by inserting a polyethylene sheet or other bond breaker between the base slab and the cellular concrete topping.

Application and Installation

Before underlayments are placed, the subfloor must be cleaned. Subfloors must be solid and free of dust and coatings like paint or wax. ASTM standards, D 4259, D 4260, D 4261, and D 4262 can be helpful in preparing concrete subfloors.

A primer is applied to a cleaned subfloor by roller, brush, sprayer, or squeegee to seal the surface and aid bond. To attain good bond—and thus enhance performance—the slab surface scheduled to receive a bonded underlayment should be prepared by scarification and/or acid etching. These preparation methods are discussed under concrete floor surface preparation beginning on page 115. For cellular concrete underlayments, the floor should be swept clean, and any hole should be filled prior to applying a bond breaker (such as kraft paper) or a bonded moisture barrier (such as liquid latex). Strip contraction joints should be installed as needed.

Depending on the product, traditional or special mixing equipment is used. Small quantities of some products can be mixed in a pail or drum using an electric drill and paddle. The product can be poured directly onto the floor from the container. On large jobs, automatic equipment is used to mix and then pump the grout to the desired location. The mixture is spread over the floor using a rake or squeegee, after which most self-leveling materials need no further finishing, as they are self-smoothing. Perimeter chalk lines and plastic guides on the subfloor help control the depth. Some coating and concrete resurfacing products are designed to be spray-applied. Metal reinforcement is usually not used or needed.

The mixture must be allowed to set and harden before further work can be done on the floor. Manufacturers can provide guidance on:

- setting and hardening time
- the time required before the product can be walked on or construction can be resumed
- when a floor covering can be applied

Chapter 9 contains information about the drying of concrete. As moisture can interfere with the performance of many floor coverings, before placing a covering, the moisture condition of the slab should be tested or a test installation should be made of the product to be used. Chapter 9 contains a description of many moisture tests for concrete floors. To avoid moisture-related problems, new full-depth concrete floors may require more than 2 months of drying before floor coverings or even some underlayments can be installed. Further information can be found in "Floor-Covering Materials and Moisture in Concrete"; see Related Publications.

CHAPTER 11

SPECIAL FLOORS

TYPES AND APPLICATIONS

Throughout this book, concrete floors on ground have been described for use in many common applications. In addition, concrete makes a good choice for floors having special technical or aesthetic needs, such as:

- white cement concrete floors
- radiant-heated floors
- cold rooms, freezers, or ice rinks
- service environments with elevated temperatures
- colored and textured floors

White Cement Concrete Floors

One of the primary uses of a white cement concrete floor is to improve illumination. Although visibility can be upgraded by increasing the number or intensity of lighting units, it costs more money to install and operate extra or more powerful lights. Light colored surfaces are beneficial because they provide maximum light reflectance. As a background, they reduce shadows from large machinery, stacked goods, and tall racks.

White cement concrete floors reflect light significantly better than gray concrete floors. Even after continued service, there is little change in light-reflecting ability with appropriate cleaning.

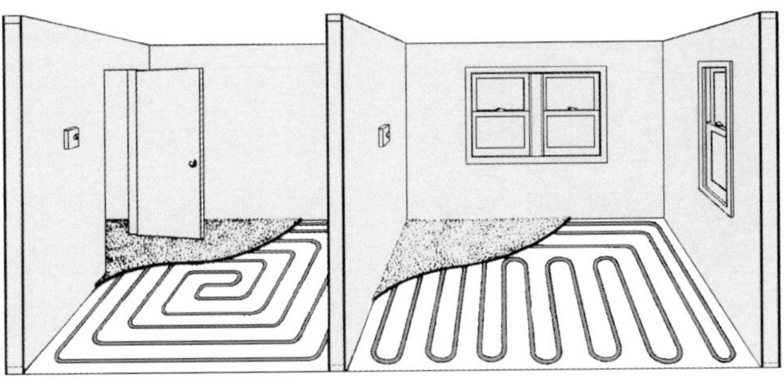

Fig. 11-1 (top left, bottom left, top right, bottom right). Concrete floors have special uses beyond the standard floors on ground (white surface, radiant heating, ice rink, foundry). (69701, 69681, K21324)

White concrete is made with white cement, water, and aggregates. White portland cement conforms to ASTM C 150, usually Type I or Type III. It is manufactured from specially selected raw materials to control the amount of iron and manganese oxides, which give cement its characteristic gray color. White ground granulated blast furnace slag conforming to ASTM C 989 can also be used as a cementitious material in white concrete.

White concrete surfaces are constructed either as toppings, full-depth floors, or white shake-on surfaces. Toppings minimize material costs while full-depth floors minimize labor costs. White concrete is well suited to two-course floor construction because the base slab can be made from gray concrete to save on the cost of materials. However, full-depth white concrete floors not only save on labor, but also the time associated with two-course construction.

Aggregate for white concrete should be light colored (or preferably, white) because dark aggregates produce darker or spotted surfaces. The aggregates should be clean so that discoloration from foreign materials is not a concern.

White mortar proportions are generally such that the mixture is about 60% fine aggregate by volume; white concrete proportions work out to be 25% to 30% fine aggregate and 50% to 55% coarse aggregate by volume.

White floors and toppings are placed and finished in the same manner as gray floors (see Fig. 11-2). Mixing and placing equipment for white concrete—mixing drums and chutes—should be clean in order to avoid discoloration. Trowels and floats should not be made of metal. Finishing tools made of fiberglass, polyethylene, or other non-marking materials prevent staining of white concrete surfaces.

White concrete should be protected during curing. Once it is hard enough to resist damage, the surface can be covered with non-staining waterproof paper. Because plastic sheeting does not always lay flat against the surface, it is more difficult to use correctly and there is an increased danger of discoloration. Curing compounds can be used if they do not interfere with the floor color.

Table 11-1. Aggregate Grading for White Concrete

Passing sieve		Percent by mass
Fine aggregate		
9.5 mm	(3/8 in.)	100
4.75 mm	(No. 4)	95-100
1.18 mm	(No. 16)	45-65
300 μm	(No. 50)	5-15
150 μm	(No. 100)	0-5
Coarse aggregate		
12.5 mm	(1/2 in.)	100
9.5 mm	(3/8 in.)	95-100
4.75 mm	(No. 4)	40-60
2.36 mm	(No. 8)	0-5

Fig. 11-2. White concrete is placed in the same manner as gray concrete. (69702)

After curing and before being put into service, white concrete floors should be cleaned. Water and a neutral soap should be scrubbed into the surface with a rotary scrubber. After rinsing and drying, a colorless liquid floor hardener or sealer can be applied to help keep the floor clean and white (Farny 2001).

Radiant-Heated Floors

Radiant floor heating has been used successfully for over 50 years. The concept is simple: warm water circulates through tubing that is buried in the floor, keeping the heat where it is needed most (see Fig. 11-3). Building occupants are then comfortable at a lower thermostat setting, which may reduce fuel costs by as much as 40%.

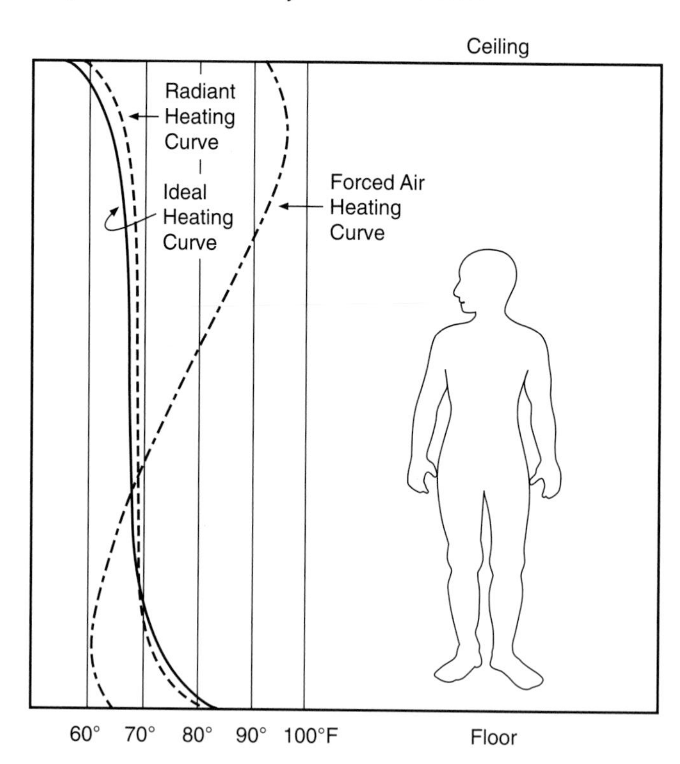

Fig. 11-3. Heat profiles from floor to ceiling show how radiant heat provides close to the ideal temperature profile for human comfort.

The biggest benefits of radiant heat are comfort and efficiency. These advantages are suited to both residential construction and commercial applications, particularly when heating spaces with large air volumes. New construction and retrofit applications are possible.

Radiant heat warms objects, not just the air. Air quality is improved since there is no hot air blowing dust and allergens throughout the building. Also, less air escapes each time windows or doors are opened.

With radiant floor heating, the heat can be zoned, or placed exactly where it is wanted and needed. For instance, in large buildings, it is possible to heat different areas to different temperatures as required by activities and occupants in that area.

Concrete is ideally suited to the benefits of radiant floor heating. Slab-on-grade applications provide an easy method of tube installation. Heated concrete slabs quickly return warmth to the room after an influx of cold air. Open doors, large windows and even high air changes will not significantly affect the building's temperature. The heat is retained in the concrete, so the room stays warmer longer without extreme temperature fluctuations.

The key to designing a radiant floor system for a concrete slab is accurate heat loss calculations. The concrete slab acts as a giant radiator to heat the building, so it is important to reduce heat losses by using insulation beneath the slab. Most radiant floor heating manufacturers have computer programs that calculate the heat loss of particular concrete slabs. These programs are very thorough: they take into account everything from water table temperatures to slab depth to floor cover R-values to determine heat loss.

Once the design is complete, installation of a radiant floor system in a concrete slab is simple. After being laid out, tubing is either stapled to the insulation or tied to the wire mesh or rebar. Fastening tubing in this way prevents it from floating to the surface during concrete placement (see Fig. 11-4). In commercial applications, tubing is often placed 300 mm (12 in.) on center. Tubing for radiant heating has an outside diameter of 13 mm to 19 mm (1/2 in. to 3/4 in.).

The following recommendations are suggested when specifying and constructing radiant-heated concrete floors:

- regular concrete mix, 21 MPa (3000 psi)
- 50-mm (2-in.) cover
- protection (no crushing) of tube
- spacing (typically 300 mm or 12 in.) and layout of tubing (zig-zag, looped, or other configuration) consistent with manufacturer's guidelines
- adequate attachment to subfloor/insulation, wire mesh, or steel reinforcement by staples or wire ties
- compliance with ASTM F 876 and F 877

The appropriate building codes should be consulted for additional requirements.

Radiant heating can be added to existing floors by embedding the tubing in a floor topping. See page 116 for information on overlaying concrete floors.

There are other applications of radiant slab heating, such as snow and ice melting. Whether for loading docks, residential driveways, city streets, sidewalks, or parking lots, concrete slabs can be kept free of ice and snow all winter (see Fig. 11-5). If the system remains under constant circulation, the pavement will remain clear and never experience the stresses caused by freeze-thaw cycles. This, and the elimination of snow removal equipment and deicing chemicals, contributes to reduced maintenance costs and a longer pavement life.

Freezer Floors and Ice Rinks

Floors that will be exposed to low temperatures during service, such as freezers, cold rooms, and ice rinks, require some special considerations. Concrete mix design and placement are standard, because the slabs are constructed

Fig. 11-4. Concrete is placed over tubing for a radiant floor. (66664)

Fig. 11-5. Compatible with all types of concrete pavements, radiant heat is useful both indoors and out. Here, streets and sidewalks are kept free of snow and ice, adding safety and increasing durability. (66665)

at normal temperatures. There are, however, at least three notable differences from standard floors:

- subgrade preparation and insulation
- thermal contraction
- potential difficulty with repairs (Garber 1991)

Cold rooms can transmit cold deep into the ground, freezing moisture in the soil and leading to subgrade heaving. To prevent this, a layer of insulation—polystyrene insulating board (styrofoam)—is installed (see Fig. 11-6). Insulation thickness commonly ranges from 100 mm to 200 mm (4 in. to 8 in.), constructed from multiple layers of 50-mm (2-in.) insulation boards. In addition, a mud slab containing or overlying heat return conduits may be installed beneath the insulation.

The modulus of subgrade reaction, or *k*-value, is dependent on insulation board density and compressive strength. Different materials provide different support, ranging from about 7 MPa/m (25 pci) on the low end to about 42 MPa/m (150 pci) on the high end. The *k*-value may be taken at an in-between value of about 21 MPa/m (75 pci) or may be determined by testing (Ringo and Anderson 1996). For large installations, the *k*-value should be determined from a mockup of the insulation system, using the modulus of subgrade reaction test (ASTM D 1196). A mockup consists of the planned thickness of insulation board over a rigid support simulating the mud slab or natural subgrade. A variation in *k* can have a significant effect on required concrete floor thickness for a given loading. Estimates made on the basis of compression-vertical strain of the insulation layer can result in serious overestimates of *k*-values and should not be used for floor thickness design.

Fig. 11-6. Freezer and other cold floors are placed on top of a layer of insulation. It is especially important to maintain form alignment during concrete placement because forms have a tendency to shift. (69613)

A second notable difference from normal temperature slabs is the amount of thermal contraction that can be expected. If a slab is constructed at 20°C (68°F), and the service temperature is about –20°C (-4°F), thermal contraction for a 100-m (330-ft) long slab will result in a shortening of about 45 mm (1.8 in.). To accommodate the shrinkage without excessive cracking, properly detailed and constructed joints and slab ends are required.

Joints will open wider in freezer slabs than in normal temperature slabs for two reasons: the thermal contraction of the concrete and the reduced friction beneath the slab. Insulation boards usually have a smooth surface and provide minimal restraint to contraction. A polyethylene slipsheet is normally installed between the insulation board top surface and the bottom of the concrete. Thermal contraction for freezer floors might be three or more times as great as for typical temperature floors. Delaying the joint filling for as long as possible allows the concrete to shrink and the joints to open. This should result in better joint filling and better joint sealant performance.

Another special consideration for freezer floors is that the floors are difficult to repair in service. Many repair materials cannot be used at low temperatures. Even mechanical methods such as grinding are difficult to accomplish inside a freezer. The difficulty and expense of removing the floor from service is much greater than it is for a typical-use concrete floor. It also takes a long time to return the floor to service, because excess moisture in the new concrete necessitates a slow reduction to operating temperature.

Ultimately, a floor for cold applications should be designed and constructed correctly the first time. This avoids difficult and potentially very costly repairs.

High-Temperature Environments

High temperatures are important in many industries, such as: iron and steel, petrochemical, nuclear, and incinerator (ACI 547 1997). Just as freezer floors have special requirements for concrete placement, floors placed in high-temperature environments must be carefully designed and constructed. Above all, they must meet all the basic requirements for concrete floors. In addition, they must meet special requirements imposed by elevated service temperatures. The structure and the material are both affected by heat.

Normal concrete that has been properly cured and dried can perform adequately when periodically subjected to concrete temperatures up to about 205°C (400°F). As long as excess mix water has been allowed time to chemically combine with the cement (by hydration) or to evaporate, there is little concern for spalling below this temperature. If desired, special aggregates can be chosen to give added protection to the concrete floor. To reduce the cost, these are sometimes added only to a topping layer rather than the entire floor thickness. All the recommendations for regular concrete toppings apply (see Chapter 8).

The most significant difference that will apply to the structural design of the floor is how the joints are handled. High temperatures will cause expansion of the concrete slab. A typical refractory concrete has a reversible thermal expansion of approximately 5×10^{-6} cm/cm /°C (3×10^{-6} in./in./°F) (ACI 547 1997). This movement should be accounted for by proper joint design. If the joints do not allow enough room for expansion, the joint faces can be subjected to pressures great enough to cause spalling and crumbling. In extreme cases, there is also potential for buckling of the slabs.

Colored and Textured Floors

Colored and textured concrete surfaces gained popularity in the 1980s and 1990s. They have widespread appeal for their looks, durability, and economy. These surface treatments are just as pleasing in the interior as they are on the exterior of a home or commercial building (see Fig. 11-7). Air entrainment is required for outdoor surfaces where freezing temperatures or deicers are anticipated. In addition to the regular benefits of concrete, these finishes enhance the prestige and value of any property. They can complete the architect's overall design and provide distinction to a project.

Pattern stamping comes alive with color. Varying styles and intensities of color can be achieved depending on whether integral pigments, dry-shake treatments, stains, or decorative aggregates are used. Aggregates, which are available in a range of colors and sizes, can complement or contrast the cement paste.

Each of the coloring techniques is briefly discussed here. Additional information on decorative concrete slabs is found in *Finishing Concrete Slabs with Color and Texture* (PA124) available from the Portland Cement Association.

Integral pigments are made from synthetic or natural metal oxides conforming to ASTM C 979, *Specification for Pigments for Integrally Colored Concrete*. Pigment is mixed into a batch of fresh concrete so that color is dispersed throughout. The dosage, typically up to 5% by mass of cement, is dependent on the intensity of color desired. Virtually any color is possible, as the use of white cement instead of gray extends the palette to include bright and pastel colors. White cement also allows for the most efficient use of pigment.

Dry-shake finishes are surface treatments used with or without pattern stamping (see Chapter 8 for more information). Colored dry-shake materials are a mixture of cement, pigments, and fine sand. A dry-shake finish is applied to the concrete surface in two or three passes, with troweling after each pass. Once the surface has been colored, it can be left to cure or it can be further worked with pattern stamping. Combining dry-shake finishes and stamping techniques allows for a wide variety of surface appearances.

Stains offer yet another method of coloring concrete (see Fig. 11-8). Unlike integral pigments or dry-shake hardeners, these materials are applied to hardened concrete. They can be used on new or old surfaces. The chemical stains react with the substrate to lock the color in. Though the color is less intense than it is with other methods, it is

Fig. 11-7. Decorative concrete floors are used in both interior and exterior applications, and are suited to high-traffic areas, such as shopping malls. (47843)

Fig. 11-8. Stained concrete makes an attractive floor for this church. (58992)

possible to create variegated effects by applying multiple stains in layers. Gritblasting patterns into the surface adds another dimension to the texture and reveals the color layers to create unusual and unique designs. Special floor waxes formulated for use with color-stained concrete not only deepen the color, but also provide an easy-to-maintain surface.

Exposed-aggregate concrete has a long history of use and continues to be popular. Exposed-aggregate finishes are rugged, slip resistant, and highly immune to wear and weather. These finishes are constructed in one of three ways:

- by seeding a select aggregate into the surface
- by mixing a chosen aggregate (usually gap-graded) into the concrete
- by applying a special topping course

A surface retarder is applied to help delay the set of the surface paste. After some setting of the concrete, aggregates are exposed by removing the surface paste with water brushing. A penetrating sealer is often applied to exposed-aggregate concrete to improve surface appearance and durability.

REFERENCES

Abrams, M. S., *Compressive Strength of Concrete at Temperatures to 1600 °F,* RD016T, Portland Cement Association, Skokie, Illinois, 1973.

Abrams, M. S., and Orals, D. L., *Concrete Drying Methods and Their Effect on Fire Resistance,* Research Department Bulletin RX181, Portland Cement Association, Skokie, Illinois, 1965.

ACPA, *Design and Construction of Joints for Concrete Highways,* American Concrete Pavement Association, Skokie, Illinois, 1991.

ACPA, *Concrete Thickness for Airport and Industrial Pavements,* MC006, American Concrete Pavement Association, Skokie, Illinois, 1992. (Formerly, Packard, R. B., Computer Program for Airport Pavement Design, SR029, Portland Cement Association, Skokie, Illinois, 1967.)

ACPA, *Recycling Concrete Pavement,* TB014, American Concrete Pavement Association, Skokie, Illinois, 1993.

ACI, Committee 301, *Standard Specifications For Structural Concrete,* ACI 301, American Concrete Institute, Farmington Hills, Michigan, 1996.

ACI, Committee 302, *Guide for Concrete Floor and Slab Construction,* ACI 302.1R, American Concrete Institute, Farmington Hills, Michigan, 1997.

ACI, Committee 306, *Cold Weather Concreting,* ACI 306R American Concrete Institute, Farmington Hills, Michigan, 1988.

ACI, Committee 318, *Building Code Requirements for Reinforced Concrete and Commentary,* ACI 318/318R, American Concrete Institute, Farmington Hills, Michigan, 1999.

ACI, Committee 504, *Joint Sealants,* ACI 504R, American Concrete Institute, Farmington Hills, Michigan, 1990 (reapproved 1997).

ACI, Committee 523, *Guide for Cast-in-Place Low-Density Concrete,* ACI 523.1R, American Concrete Institute, Farmington Hills, Michigan, 1992.

ACI, Committee 523, *Guide for Cellular Concrete Above 50 pcf, and for Aggregate Concretes Above 50 pcf with Compressive Strengths Less Than 2500 psi,* ACI 523.3R, American Concrete Institute, Farmington Hills, Michigan, 1993.

ACI, Committee 544, *State-of-the-Art Report on Fiber Reinforced Concrete,* ACI 544.1R, American Concrete Institute, Farmington Hills, Michigan, 1997.

ACI, Committee 547, *Refractory Concrete: Abstract of State-of-the-Art Report,* ACI 547R-79, American Concrete Institute, Farmington Hills, Michigan, 1997.

Basham, Kim, "Fine Grading with Tractor Loaders," *Concrete Construction,* Hanley-Wood, LLC, Addison, Illinois, May 1998, pages 64–65.

Brinckerhoff, C. H., *Report to ASTM C-9 Subcommittee III-M (Testing of Concrete for Abrasion) Cooperative Abrasion Test Program,* University of California and Portland Cement Association, Skokie, Illinois, 1970.

Bureau of Reclamation, *Concrete Manual,* 8th ed., U.S. Bureau of Reclamation, Denver, 1975, page 11.

Butt, Thomas K. "Avoiding and Repairing Moisture Problems in Slabs on Grade," *The Construction Specifier,* Alexandria, Virginia, December 1992.

CCA, *Concrete Industrial Floor and Pavement Design,* Cement and Concrete Association of Australia, North Sydney, New South Wales, 1985, 34 pages.

Chaplin, R.G., "The Influence of Cement Replacement Materials, Fine Aggregate and Curing on Abrasion Resistance of Concrete Floor Slabs," *Cement and Concrete Association,* Wexham Springs, England, 1986, page 11.

Childs, L.D., *Tests of Concrete Pavement Slabs on Gravel Subbases, Tests of Concrete Pavements on Crushed Stone Subbases, Tests of Concrete Pavement Slabs on Cement-Treated Subbases,* Development Department Bulletins, respectively: DX21, 1958; DX65, 1963; DX86, Portland Cement Association, Skokie, Illinois, 1964.

Colley, B.E. and Humphrey, H.A., *Aggregate Interlock at Joints in Concrete Pavements,* Development Department Bulletin DX124, Portland Cement Association, Skokie, Illinois, 1967.

CSA, *Concrete Materials and Methods of Concrete Construction,* A23.1-94, Canadian Standards Association, Toronto, Ontario, Canada, June 1994.

EPA, *Model Standards and Techniques for Control of Radon in New Residential Buildings,* EPA 402-R-94-009, United States Environmental Protection Agency, March 1994.

Farny, J.A., *White Cement Concrete,* EB217, Portland Cement Association, Skokie, Illinois, 2001.

Farny, J.A., and Kosmatka, S.H., *Diagnosis and Control of Alkali-Aggregate Reactions in Concrete,* IS413, Portland Cement Association, Skokie, Illinois, 1997.

Friberg, B.F., "Frictional Resistance Under Concrete Pavements and Restraint Stresses in Long Reinforced Slabs," *Proceedings,* Highway Research Board, 1954, pages 167–184.

Garber, G., *Design and Construction of Concrete Floors,* Halsted Press, a division of John Wiley & Sons, Inc., New York, 1991.

Gaul, R. W., "Moisture-Caused Coating Failures: Facts and Fiction," *Concrete Repair Digest,* Hanley-Wood, LLC, Addison, Illinois, October/November 1996, pages 255 –259.

Gebler, S. H., Klieger, P., *Effect of Fly Ash on Some of the Physical Properties of Concrete,* Research Department Bulletin RD089, Portland Cement Association, Skokie, Illinois, 1985.

Gilbert, R. I., "Shrinkage Cracking in Fully Restrained Concrete Members," *ACI Structural Journal,* American Concrete Institute, Farmington Hills, Michigan, March-April 1992, pages 141–49.

Greening, N. R., and Landgren, R., *Surface Discoloration of Concrete Flatwork,* Research Department Bulletin RX203, Portland Cement Association, Skokie, Illinois, 1966.

Goodyear, *Off-The-Road Tires Engineering Data Book,* Goodyear Tire and Rubber Co., Akron, Ohio, 2001, (view at www.goodyearotr.com/otr/index.html).

Goeb, E. O., "Technical Talk: Do Plastic Fibers Replace Wire Mesh in a Slab on Grade?" *Concrete Products,* Primedia Intertec, Overland Park, Kansas, March 1989, page 11.

Haddadin, M.J., *Computer Program for the Analysis and Design of Foundation Mats and Combined Footings,* SR100, Portland Cement Association, Skokie, Illinois, 1971.

Hanley-Wood, "Effect of High Temperature on Hardened Concrete," *Concrete Construction,* Hanley-Wood, LLC, Addison, Illinois, November 1971, pages 477–479.

Hanson, J.A., *Effects of Curing and Drying Environments on Splitting Tensile Strength of Concrete,* Development Department Bulletin DX141, Portland Cement Association, Skokie, Illinois, 1968.

Hedenblad, G., *Drying of Construction Water in Concrete–Drying Times and Moisture Measurement,* T9, Swedish Council for Building Research, Stockholm, 1997. [Available from Portland Cement Association as LT229.]

Hedenblad, G., "Drying of Construction Water in Concrete–The Swedish Experience," *Issues in Moisture Migration in Concrete,* ACBM Technology Transfer Day, NSF Center for Advanced Cement-Based Materials (at Northwestern University), Evanston, Illinois, April 1998.

Hetenyi, M., *Beams on Elastic Foundations,* The University of Michigan Press, Ann Arbor, Michigan, 1946, 255 pages.

HRB, *Road Test One-MD,* Highway Research Board, 1952.

HRB, *The AASHTO Road Test,* Highway Research Board Special Report No. 61E, 1962.

Kanare, H. M., *Understanding Concrete Floors and Moisture Issues,* CD014, Portland Cement Association, Skokie, Illinois, 2000.

Kelley, E.F., "Applications of the Results of Research to the Structural Design of Concrete Pavements," *Public Roads,* Vol. 20, No. 5, July 1939, pages 83–104.

Kosmatka, S. H., "Floor-Covering Materials and Moisture in Concrete," *Concrete Technology Today,* PL853, Portland Cement Association, Skokie, Illinois, September 1985.

Kosmatka, S. H., "Sulfate-Resistant Concrete," *Concrete Technology Today,* PL883, Portland Cement Association, Skokie, Illinois, October 1988, 3 pages.

Kosmatka, S. H., "Bleeding," *Significance of Tests and Properties of Concrete and Concrete-Making Materials,* STP 169C, American Society for Testing and Materials, West Conshohocken, Pennsylvania, 1994, pages 88–111. [Also available from PCA as RP328.]

Kosmatka, S. H., and Panarese, W.C., *Design and Control of Concrete Mixtures,* EB001, 13th edition, Portland Cement Association, Skokie, Illinois, revised 1994, 212 pages.

Kunt, M. M., and McCullogh, B. F., "Evaluation of the Subbase Drag Formula by Considering Realistic Subbase Friction Values," *Transportation Research Record 1286,* Transportation Research Board National Research Council Washington, D.C., 1990, pages 78–83.

Landgren, R., and Hadley, D.W., *Surface Popouts Caused by Alkali-Aggregation Reaction,* RD121, Portland Cement Association, Skokie, Illinois, (to be published) late 2001.

Liu, T. C., "Abrasion Resistance of Concrete," *ACI Journal,* American Concrete Institute, Farmington Hills, Michigan, September–October 1981.

Lytton, R.L., and Meyer, K.T., "Stiffened Mats on Expansive Clay," *Journal,* Soil Mechanics and Foundations Division, American Society of Civil Engineers, July 1971.

Marais, L. R., and Perrie, B. D., *Concrete Industrial Floors on the Ground,* Portland Cement Institute, Midrand, South Africa, 1993.

Miller, F.M., Powers, L.J., Taylor, P.C., *Investigation of Discoloration in Concrete Slabs,* Serial No. 2228, Portland Cement Association, Skokie, Illinois, 1999, 22 pages.

Monfore, G. E., *A Small Probe-Type Gage for Measuring Relative Humidity,* Research Department Bulletin RX160, Portland Cement Association, Skokie, Illinois, May 1963.

NAHB, *Building Radon Resistant Foundations,* #TR104, National Association of Home Builders National Research Center, Upper Marlboro, Maryland, 1989.

NAS, *Critical for Selection and Design of Residential Slabs-on-Ground,* U.S. National Academy of Sciences Publication 1571, Washington, D.C., 1968.

Nicholson, L. P., "How to Minimize Cracking and Increase Strength of Slabs on Grade," *Concrete Construction,* Hanley-Wood, LLC, Addison, Illinois, 1981, pages 739–742.

Nowlen, W. J., "Influence of Aggregate Properties on Effectiveness of Interlock Joints in Concrete Pavements," *Journal of the PCA Research and Development Laboratories,* Vol. 10, No. 2, Portland Cement Association, Skokie, Illinois, May 1968.

NTMA, *Terrazzo Specifications and Design Guide,* The National Terrazzo and Mosaic Association, Inc., Leesburg, Virginia, 1999, 50 pages (pdf).

Okamoto, P. A., and Nussbaum, Peter J., *Concentrated Loads on Industrial Concrete Floors on Grade,* Serial No. 1943, Portland Cement Association, Skokie, Illinois, July 1984.

Packard, R.G., *Computer Program for Airport Pavement Design,* SR029, Portland Cement Association, 1967.

Packard, R.G., and Spears, Ralph E., *Slab Thickness Design for Factory or Warehouse Floors,* Serial No. 1298, Portland Cement Association.

PCA 1958, "Tentative Recommendations for Prestressed Concrete," *Journal of the American Concrete Institute,* v.29, no.7, Jan. 1958, Proceedings, v.54, Portland Cement Association, Skokie, Illinois, January.

PCA 1966, *Thickness Design for Concrete Pavements,* IS010P, Portland Cement Association, Skokie, Illinois.

PCA 1985, "Popouts: Causes, Prevention, Repair," *Concrete Technology Today,* PL852, Portland Cement Association, Skokie, Illinois, June.

PCA 1991, *Fiber Reinforced Concrete,* SP039, Portland Cement Association, Skokie, Illinois, September.

PCA 1992, *Painting Concrete,* IS134, Portland Cement Association, Skokie, Illinois.

PCA 1995a , "Cleaning Concrete Pavements by Power Washing," *Concrete Technology Today,* PL953, Portland Cement Association, Skokie, Illinois, November.

PCA 1995b , "Early Sawing to Control Slab Cracking," *Concrete Technology Today,* PL953, Portland Cement Association, Skokie, Illinois, November.

PCA 1995c, *PCA-MATS—Analysis and Design of Mat Foundations,* Combined Footings, and Slabs on Grade, MC012 computer program, Portland Cement Association, Skokie, Illinois.

PCA 1996a, *Resurfacing Concrete Floors,* IS144, Portland Cement Association, Skokie, Illinois.

PCA 1996b, "Specifying a Burnished Floor Finish," *Concrete Technology Today,* PL963, Portland Cement Association, Skokie, Illinois, December.

PCA 1997, "Radiant Heat with Concrete," *Concrete Technology Today,* PL971, Portland Cement Association, Skokie, Illinois, April.

PCA 2001, *Effects of Substances on Concrete and Guide to Protective Treatments,* IS001, Portland Cement Association, Skokie, Illinois.

Pickett, G., and Ray, G.K., "Influence Charts for Concrete Pavements," American Society of Civil Engineers, *Transactions,* Paper No. 2425, Vol. 116, 1951, pages 49–73.

PTI, *Construction and Maintenance Procedures Manual for Post-Tensioned Slabs-on-Ground,* Second Edition, Post-Tensioning Institute, Phoenix, Arizona, 1998, 84 pages.

PTI, *Design and Construction of Post-Tensioned Slabs-on-Ground,* Second Edition, Post-Tensioning Institute, Phoenix, Arizona, 1996, 90 pages.

PTI, *Post-Tensioned Commercial and Industrial Floors,* Post-Tensioning Institute, Phoenix, Arizona, 1983.

PTI, *Post-Tensioning Manual,* Fifth Edition, Post-Tensioning Institute, Phoenix, Arizona, 1990.

Ray, G.K., Cawley, M.L., and Packard, R.G., "Concrete Airport Pavement Design—Where Are We?" *Airports, Key to the Air Transportation System,* American Society of Civil Engineers, April 1971, pages 183–226.

Rice, P.F., "Design of Concrete Floors on Ground for Warehouse Loadings," Title No. 54-7, *ACI Journal,* Vol. 29, No. 2, August 1957, pages 105–113.

Ringo, B., *Reinforcing Steel in Slabs-on-Grade,* Engineering Data Report No. 37, Concrete Reinforcing Steel Institute (and Tech Facts TF701, Wire Reinforcement Institute), Schaumburg, Illinois, 1991.

Ringo, B., and Anderson, R. B., *Designing Floor Slabs on Grade,* Hanley-Wood, LLC, Addison, Illinois, 1996.

Schrader, E. K., "Square Dowels Control Slab Curling," *Concrete Construction,* Hanley-Wood, LLC, Addison, Illinois, 1999.

Suprenant, B. A., 1998a, "Qualifying Quick-Dry Concrete," *Concrete Producer,* Hanley-Wood, LLC, Addison, Illinois, September, pages 619–620.

Suprenant, B. A., 1998b, "Quick-Dry Concrete: A New Market for Ready-Mix Producers," *Concrete Producer,* Hanley-Wood, LLC, Addison, Illinois, May, pages 330–333.

Suprenant, B. A., and Malisch, W. R., "Where to Place the Vapor Retarder," *Concrete Construction,* Hanley-Wood, LLC, Addison, Illinois, May 1998, pages 427–433.

Tabatabuie, A. M., Barenberg, E. J., and Smith, R. E., "Longitudinal Joint Systems in Slip-Formed Rigid Pavements, Volume II, Analysis of Load Transfer Systems for Concrete Pavements, *Report No. FAA-RD 79-4,* II, Federal Aviation Administration, Washington D. C., November 1979.

Taylor, P.C., Detwiler, R.J., and Tang, F.J., *Investigation of Discoloration of Concrete Slabs (Phase 2),* Serial No. 2228b, Portland Cement Association, Skokie, Illinois, 2000, 22 pages.

The Tire and Rim Association, *1974 Yearbook,* Akron, Ohio, 1974.

Transtec, *HIPERPAV™ Computer Program,* Federal Highway Administration, U.S. Department of Transportation FHWA Research Development and Technology, Turner-Fairbank Highway Research Center, 6300 Georgetown Pike, McClean Virginia, 22101-2296, October 1999.

Walker, W. W., Holland, J. A., "Plate Dowels for Slabs on Ground," *Concrete International,* American Concrete Institute, Farmington Hills, Michigan, July 1998, pages 32–35.

Wambold, J. C., Ph.D., and Antle, C. E., Ph.D., *Evaluation of F-Number System and Waviness Index for Measuring Floor Flatness and Levelness,* American Concrete Institute, Farmington Hills, Michigan, 1996, 10 pages.

Westergaard, H.M., "Computation of Stresses in Concrete Roads," *Proceedings,* Highway Research Board, Vol. 5, Part I, 1925, pages 90–112.

Wood, S. L., *Evaluation of the Long-Term Properties of Concrete,* RD102T, Portland Cement Association, Skokie, Illinois, 1992.

Wu, C., and Okamoto, P. A., *Refinement of the Portland Cement Association Thickness Design Procedure for Concrete Highway and Street Pavements,* Serial No. 1916, Portland Cement Association, Skokie, Illinois, 1992

STANDARDS—ASTM AND OTHER

ASTM A 82-97a, *Standard Specification for Steel Wire, Plain, for Concrete Reinforcement*

ASTM A 184/A 184M, *Standard Specification for Fabricated Deformed Steel Bar Mats for Concrete Reinforcement*

ASTM A 185, *Standard Specification for Steel Welded Wire Fabric, Plain, for Concrete Reinforcement*

ASTM A 496, *Standard Specification for Steel Wire, Deformed, for Concrete Reinforcement*

ASTM A 497, *Standard Specification for Steel Welded Wire Fabric, Deformed, for Concrete Reinforcement*

ASTM A 615/A 615M, *Standard Specification for Deformed and Plain Billet-Steel Bars for Concrete Reinforcement*

ASTM A 704/A 704M, *Standard Specification for Welded Steel Plain Bar or Rod Mats for Concrete Reinforcement*

ASTM A 706/A 706M, *Standard Specification for Low-Alloy Steel Deformed and Plain Bars for Concrete Reinforcement*

ASTM A 767/A 767M, *Standard Specification for Zinc-Coated (Galvanized) Steel Bars for Concrete Reinforcement*

ASTM A 820, *Specification for Steel Fibers for Fiber-Reinforced Concrete*

ASTM A 884/A 884M, *Specification for Epoxy-Coated Steel Wire and Welded Wire Fabric for Reinforcement*

ASTM A 955M, *Standard Specification for Deformed and Plain Stainless Steel Bars For Concrete Reinforcement [Metric]*

ASTM A996/A996M-01, *Specification for Rail-Steel and Axle-Steel Deformed Bars for Concrete Reinforcement*

ASTM C 33, *Specification for Concrete Aggregates*

ASTM C 42, *Method for Obtaining and Testing Drilled Cores and Sawed Beams of Concrete*

ASTM C 78, *Test Method for Flexural Strength of Concrete (Using Simple Beam with Third-Point Loading)*

ASTM C 94, *Standard Specification for Ready-Mixed Concrete*

ASTM C 150, *Standard Specification for Portland Cement*

ASTM C 157, *Standard Test Method for Length Change of Hardened Hydraulic-Cement Mortar and Concrete*

ASTM C 174, *Standard Test Method for Measuring Thickness of Concrete Elements Using Drilled Concrete Cores*

ASTM C 232, *Test Methods for Bleeding of Concrete*

ASTM C 260, *Specification for Air-Entraining Admixtures for Concrete*

ASTM C 295, *Practice for Petrographic Examination of Aggregates for Concrete*

ASTM C 330, *Specification for Lightweight Aggregates for Structural Concrete*

ASTM C 494 and ASTM C 494M, *Specification for Chemical Admixtures for Concrete*

ASTM C 595, *Standard Specification for Blended Hydraulic Cements*

ASTM C 597, *Test Method for Pulse Velocity Through Concrete*

ASTM C 618, *Specification for Coal Fly Ash and Raw or Calcined Natural Pozzolan for Use as a Mineral Admixture in Portland Cement Concrete*

ASTM C 779, *Test Method for Abrasion Resistance of Horizontal Concrete Surfaces*

ASTM C 856, *Practice for Petrographic Examination of Hardened Concrete*

ASTM C 881, *Specification for Epoxy-Resin-Base Bonding Systems for Concrete*

ASTM C 882, *Test for Bond Strength of Epoxy-Resin Systems Used with Concrete*

ASTM C 979, *Specification for Pigments for Integrally Colored Concrete*

ASTM C 989, *Specification for Ground Granulated Blast-Furnace Slag for Use in Concrete and Mortars*

ASTM C 1012, *Test Method for Length Change of Hydraulic-Cement Mortars Exposed to a Sulfate Solution*

ASTM C 1017 and ASTM C 1017M, *Specification for Chemical Admixtures for Use in Producing Flowing Concrete*

ASTM C 1116, *Specification for Fiber-Reinforced Concrete and Shotcrete*

ASTM C 1138, *Test Method for Abrasion Resistance of Concrete (Underwater Method)*

ASTM C 1157, *Standard Performance Specification for Blended Hydraulic Cement*

ASTM C 1240, *Specification for Use of Silica Fume as a Mineral Admixture in Hydraulic-Cement Concrete, Mortar, and Grout*

ASTM C 1404, *Standard Test Method for Bond Strength of Adhesive Systems Used with Concrete as Measured by Direct Tension*

ASTM D 427, *Test Method for Shrinkage Factor of Soils*

ASTM D 698, *Test Method for Laboratory Compaction Characteristics of Soil Using Standard Effort (12,400 ft-lbf/ft^3 (600 kN-m/m^3))*—also known as the "Standard Proctor Test"

ASTM D 1194, *Test Method for Bearing Capacity of Soil for Static Load and Spread Footings*

ASTM D 1196, *Methods for Nonrepetitive Static Plate Load Tests of Soils and Flexible Pavement Components, for Use in Evaluation and Design of Airport and Highway Pavements*

ASTM D 1557, *Test Method for Laboratory Compaction Characteristics of Soil Using Modified Effort (56,000 ft-lbf/ft^3 (2700 kN-m/m^3))*—also known as the "Modified Proctor Test"

ASTM D 1883, *Test Method for CBR (California Bearing Ratio) of Laboratory-Compacted Soils*

ASTM D 2240, *Test Method for Rubber Property—Durometer Hardness*

ASTM D 2487, *Practice for Classification of Soils for Engineering Purposes (Unified Soils Classification System)*

ASTM D 4259, *Practice for Abrading Concrete*

ASTM D 4260, *Practice for Acid Etching Concrete*

ASTM D 4261, *Practice for Surface Cleaning Concrete Unit Masonry for Coating*

ASTM D 4262, *Test Method for pH of Chemically Cleaned or Etched Concrete Surfaces*

ASTM D 4263, *Test Method for Indicating Moisture in Concrete by the Plastic Sheet Method*

ASTM D 4318, *Test Method for Liquid Limit, Plastic Limit, and Plasticity Index of Soils*

ASTM D 4397, *Specification for Polyethylene Sheeting for Construction, Industrial, and Agricultural Applications*

ASTM D 4429, *Test Method for CBR (California Bearing Ratio) of Soils in Place*

ASTM D 4580, *Practice for Measuring Delaminations in Concrete Bridge by Sounding*

ASTM D 4633, *Test Method for Stress Wave Energy Measurement for Dynamic Penetrometer Testing Systems*

ASTM E 96, *Test Methods for Water Vapor Transmission of Materials*

ASTM E 119, *Test Methods for Fire Tests of Building Construction and Materials*

ASTM E 1155, *Test Method for Determining Floor Flatness and Levelness Using the F-Number System (Inch-Pound Units)*

ASTM E 1155 M, Test Method for Determining Floor Flatness and Levelness Using the F-Number System (Metric Units)

ASTM E 1465, *Standard Guide for Radon Control Options for the Design and Construction of New Low Rise Residential Buildings*

ASTM E 1486, *Standard Test Method for Determining Floor Flatness and Levelness Using Waviness, Wheel Path and Levelness Criteria*

ASTM E 1486 M, *Standard Test Method for Determining Floor Flatness and Levelness Using Waviness, Wheel Path and Levelness Criteria (Metric Units)* Check metric title

ASTM E 1643, *Practice for Installation of Water Vapor Retarders Used in Contact with Earth or Granular Fill Under Concrete Slabs*

ASTM E 1745, *Specification for Plastic Water Vapor Retarders Used in Contact with Soil or Granular Fill Under Concrete Slabs*

ASTM F 876, *Specification for Crosslinked Polyethylene (PEX) Tubing*

ASTM F 877, *Specification for Crosslinked Polyethylene (PEX) Plastic Hot- and Cold-Water Distribution Systems*

ASTM F 1869, *Test Method for Measuring Moisture Vapor Emission Rate of Concrete Subfloor Using Anhydrous Calcium Chloride*

CAN/CSA Standard B137.5M-96, *Crosslinked Polyethylene (PEX) Tubing Systems for Pressure Applications*

CAN/CSA-A5/A8/A362-93, *Portland Cement/Masonry Cement/Blended Hydraulic Cement*

CAN/CSA A 23.1-94, *Concrete Materials and Methods of Concrete Construction*

CAN/CSA A 23.5-M86 (R1992), *Supplementary Cementing Materials*

CAN/CSA A 363-93, *Cementitious Hydraulic Slag*

ANSI/NSF Certification 14 (Ingredients, materials, products, quality assurance, and marking for thermoplastic and thermoset piping systems)

ANSI/NSF Certification 61 (Toxicity for plumbing materials)

ICBO Evaluation Report No. 4407 (Piping with oxygen diffusion barrier)

ICBO Evaluation Report No. 5142 (Piping with no oxygen diffusion barrier)

GLOSSARY

abrasion resistance: the wear resistance of a concrete floor surface as measured according to ASTM C 944.

aggregate interlock: load transfer mobilized at a control (contraction) joint due to contact between aggregate at the crack surfaces located below control joint when one slab edge deflects under a load.

Atterberg limits: see plasticity.

axle load: the amount of weight supported by the front axle of a lift truck with a load placed on the front forks (used for floor thickness design).

backer rod: a leave-in insert placed at the bottom of a sealant reservoir to create a convex shape at the sealant bottom in order to maximize sealant extensibility; backer rods are used for pavements exposed to pneumatic tire traffic and should not be installed in slab-on-grade floors at interior locations and/or slabs exposed to solid tire or caster traffic.

bleeding: settlement of the heavier particles (cement, aggregate) in fresh concrete that displace mix air and water upward toward the slab surface before hardening.

blisters: raised surface bumps in the concrete surface due to premature densification that traps bleed air or water beneath the surface.

bushhammer: a tool used to scarify concrete surfaces—sometimes used for concrete repair (to remove unsound material).

butt joint: the location where the vertical edges of two adjacent slabs meet; see construction joint.

compressive strength: a measure of the maximum load supported per unit area (failure strength) of an unconfined cylindrical soil (ASTM D 2166) or concrete (ASTM C 39) specimen.

construction joint: a vertical surface that meets at predetermined or unplanned (emergency) locations to facilitate concrete placement sequences; concrete in the first placement hardens before starting adjacent floor placement.

control or contraction joint: a sawed or formed groove in a slab that steers cracks to align at predetermined locations (beneath the groove) in order to minimize (control) random slab cracking.

continuously reinforced floor: a concrete slab that contains deformed bar reinforcement (0.5% to 0.7% of cross section area); the slab has no control joints because the reinforcement induces closely spaced tight transverse cracks.

contraction joint: see control joint.

crazing, crazing cracks: small pattern cracks occurring in a slab surface, associated with early surface drying or cooling; crazing cracks are generally less than 6 mm (1/4 in.) deep and are generally not structurally significant.

curing: maintaining adequate moisture and temperature in freshly placed concrete to optimize early strength development and durability.

curing membrane: a coating or cover on the concrete surface to prevent evaporation of moisture.

curling: upward deformations that occur at slab edges and corners due to shrinkage of concrete near the slab surface; cooler and drier conditions at the surface increase curling; see warping.

delamination: an area of surface mortar, generally about 3 to 5 mm (1/8 to 1/4 in.) thick, that has separated from underlying concrete (roughly horizontal layer); it results from a prematurely closed (densified) surface that prevents bleed water and air from escaping.

density: the measure of soil subgrade and subbase compaction, as a percentage of compaction attainable, as measured according to ASTM D 698.

dowel: a short length of smooth steel bar installed across joints; smooth bars (no deformations) provide load transfer that keeps slabs vertically aligned under load while permitting the joint to open and close as needed for concrete volume changes (expansion or contraction).

drag: the resistance to free horizontal movement provided by subgrade friction when concrete slabs change volume (shrink).

dry-shake: surface application of a mixture of dry cement and other materials; these materials can be hard fine mineral or metallic aggregate to improve resistance to abrasion or can be fine aggregate mixed with a pigment to color the slab.

drying shrinkage: decrease in length (and volume) of hardened concrete element due to loss of moisture.

dusting: the development of a weak, powdery material on the concrete surface; often due to improper concreting practices (placing or curing), poor wear resistance, or abrasive traffic.

effective prestress: the amount of compressive stress that remains in a post-tensioned slab after accounting for losses from subgrade friction, strand friction, concrete creep, strand relaxation, elastic concrete shortening, and strand wobble friction.

entrained air: the microscopic air voids purposely introduced into the concrete during mixing by means of a chemical admixture to improve fresh or hardened concrete properties.

entrapped air: air voids that occur in fresh and hardened concrete as a result of concrete mixing and placing—usually no more than 1% to 2.5% of concrete volume; entrapped air voids are larger in size than entrained air voids.

expansion joint: see isolation joint.

factor of safety: the ratio of flexural strength (modulus of rupture) of the concrete to stress due to loading of a slab-on-grade or the ratio of flexural strength to the design working stress.

fatigue failure: the cracking of a concrete slab due to repeated loadings.

finishing: the construction activities following concrete deposition on grade starting with strikeoff and completed with broom texturing or hard troweling, depending on surface finish.

flatness: the measure of smoothness of a concrete floor surface as determined by ASTM E 1155 or by the 3 m (10 ft) straightedge; a distinction is made between inclined or sloped surfaces and level surfaces—the inclined surfaces have flatness but not levelness.

flexural strength: a measure of concrete's resistance to bending, as tested according to ASTM C 78; frequently used in floor and pavement design.

floating: the finishing technique following bullfloating and air and/or water bleeding. this step compacts the surface, slightly embeds coarse aggregate, and removes small surface imperfections; it levels the surface and consolidates the mortar for further finishing.

fluosilicate: a zinc or magnesium fluoride used as a floor hardener.

grout: a mixture of sand with cement and water (sand-cement grout) or a mixture of only cement and water (neat grout) used as a bonding agent between an existing slab and a concrete topping.

isolation joint: a through-slab separation that allows the two adjacent members to move freely of one another, both horizontally and vertically; these are used at edges of floor slabs or around inserts within the floor such as machine foundations, column bases, etc.

joint: a separation through a concrete section that is used for controlling cracking and for facilitating concrete placement; there are three common types of joints: control or contraction, construction, and isolation (expansion). see specific type for definition.

joint sealant: material placed into joint reservoirs to prevent ingress of foreign objects and, if needed, to provide support to adjacent vertical faces of slab edges.

joint sealant reservoir: the space formed (generally by sawing) at the upper portion of slab edges to accommodate joint sealant(s).

liquid limit: lower limit of viscous flow of a soil; a measure of the water content at which a soil exhibits a change in behavior from a workable (plastic) state to a flowing (fluid) state.

levelness: a measure of how close to a horizontal plane or how different from a horizontal plane the overall surface of a floor is, as measured by ASTM E 1155.

membrane curing: the prevention of water evaporation from a fresh concrete surface by application of a liquid curing compound, plastic, or paper sheets.

mesh, mesh reinforcement: see welded wire fabric.

modulus of elasticity: the ratio of concrete stress to strain either determined by flexural strength or compressive strength testing (ASTM C 469).

modulus of rupture: see flexural strength.

modulus of subgrade reaction: the measure of the amount of support provided by the material on which the concrete slab rests, as determined by a plate bearing test according to ASTM D 1195.

moisture retarder: see vapor retarder.

mortar: a mixture of cement, fine aggregate, and water.

plastic index: lower limit of the plastic state of a soil; a measure of the water content at which a soil exhibits a change in behavior from a workable (plastic) state to a non-workable (solid) state.

plasticity: the measure of deformability of a fine grained (clayey) soil as determined in accordance with ASTM D 4318; the difference between the liquid limit and the plastic limit of a soil.

post-tensioning: the application of a tensile stress to a steel tendon or bar, balancing the force with the compressive strength of the concrete floor.

prestressing: to apply a compressive stress to the concrete floor slab by post-tensioning. see post-tensioning.

raveling: tearing of aggregate at sawcut control joint edges occurring due to insufficient paste-aggregate bond strength at time of jointing.

reinforcement: the deformed round steel bars or welded wire fabric located in the concrete slab to carry tensile forces across cracks; in structurally reinforced slabs-on-grade, the reinforcement is proportioned to work together with the concrete to carry the imposed loads in bending.

restraint stress: the force (tensile, compressive, bending) that is developed in a concrete member as a result of the concrete being held tightly in place while it is being subjected to loads or changes in moisture or temperature; the forces that are acting would lead to movement or a change in volume if the concrete were not held in place.

safety factor: see factor of safety.

sawed joint: control joint reservoir and/or weakened plane made by sawing with a diamond or carborundum blade.

scaling: the process of surface mortar loss by flaking that exposes coarse aggregate below the surface; depending on severity, can also include loss of near-surface coarse aggregate; usually occurs on inadequately air-entrained concrete that is exposed to freezing and thawing and, possibly, deicers.

screeding: the first activity immediately following concrete placement in forms that is the initial step in shaping the floor surface; also called strikeoff.

shrinkage compensating concrete: a concrete mix that expands during curing. this expansion is transferred to the reinforcement, which is put into tension. the overall effect is "shrinkage compensation," which allows construction of large floor slabs having fewer joints without intermediate cracking.

shrinkage limit: the maximum water content of a cohesive soil at which no volume change will occur as the soil dries; measured in accordance with ASTM D 4943.

slab stabilization: the introduction of a grout mix into the void space between slab bottom and subbase/subgrade to restore uniform slab support.

slabjacking: a special version of slab stabilization used to return slabs to proper elevation. pressurized grout is pumped under the slab and the surface elevation is monitored.

slump: a measure of the consistency of fresh concrete as determined by ASTM C 143; values are reported in mm (0 to 300) or in inches (0 to 12).

spalling: the breaking off of concrete edges at joint faces occurring due to solid tire or caster traffic when adequate constraint to the concrete is not provided by joint sealant.

strain: change in length due to load, reported as deformation per length of original member: dimensionless length change, mm per mm (in. per in.).

stress: force per unit area, reported as MPa or psi; can be tension, flexure (bending), or compression.

subbase: the (optional) granular or stabilized soil layer located on top of the subgrade and beneath the slab bottom.

subgrade: the soil located below the (optional) subbase or concrete slab bottom; the subgrade soil can be either an imported and compacted fill or the existing soil after stripping unsuitable materials and compacting the top 150 mm to 300 mm (6 in. to 12 in.).

subgrade friction: the measure of resistance to horizontal slab movement provided by the subbase or subgrade.

thermal contraction: decrease in length due to cooling.

troweling: the finishing activity that follows floating; it densifies the slab surface and improves surface flatness.

vapor retarder: a low-permeability membrane installed below a concrete slab-on-grade floor to minimize moisture vapor migration through the slab.

vibration: the act of consolidating granular soils or fresh concrete by applying repeated high frequency mechanical shaking to reduce the size and number of voids between particles.

warping: upward lifting that occurs at slab edges and corners due to concrete near the slab surface being drier or cooler than concrete at greater depth in the slab; see curling.

water-cement ratio: the proportion by weight of water to cement in a concrete, mortar, or grout mix.

water reducer: a chemical admixture that reduces the amount of water required for a certain workability in fresh concrete.

wear resistance: see abrasion resistance.

welded wire fabric, welded wire mesh: steel wires welded into a two-dimensional grid for use as reinforcement to hold cracks tight in a concrete slab.

working stress: the design stress obtained by the ratio of maximum allowable stress (modulus of rupture) to the factor of safety.

METRIC CONVERSIONS

Following are metric conversions of the measurements used in this text. They are based in most cases on the International System of Units (SI).

1 in.	=	25.40 mm
1 sq. in.	=	645.16 mm^2
1 ft	=	0.305 m
1 sq ft	=	0.093 m^2
1 sq ft per gallon	=	0.025m^2/L
1 gal	=	3.8 L
1 kip	=	4.5 kN
1 lb	=	0.454 kg
1 lb per cubic yard	=	0.5993 kg/m^3
1 pci (or psi/in.)	=	0.27 MPa/m
1 psf	=	4.88 kg/m^2
1 psi	=	0.0069 MPa
No. 4 sieve	=	4.75 mm
No. 200 sieve	=	75 μm
1 bag cement (U.S.)	=	94 lb = 42.7 kg
1 bag cement (Canadian)	=	88 lb = 40 kg
1 bag per cubic yard (U.S.)	=	55.8 kg/m^3
For permeance:		
1 perm	=	1 grain/h-ft^2 in. Hg
	=	0.659 metric perm

RELATED PUBLICATIONS

The publications cited in this text as well as other related publications can be purchased from Portland Cement Association. The following are particularly useful:

Cement Mason's Guide, (PA122)

Concrete Slab Surface Causes, Prevention, Repair, (IS177)

Design and Control of Concrete Mixtures, (EB001)

"Discoloration of Concrete—Causes and Remedies," *Concrete Technology Today* (PL861)

Drying of Construction Water in Concrete—Drying Times and Moisture Measurement (LT229)

Effects of Substances on Concrete and Guide to Protective Treatments (IS001)

"Floor-Covering Materials and Moisture in Concrete," *Concrete Technology Today* (PL883)

"How to Double the Value of Your Concrete [Floor] Dollar," *Concrete Technology Today* (PL852)

Maintenance of Joints and Cracks in Concrete Pavements (IS188)

Painting Concrete (EB007)

PCA *Soil Primer* (EB007)

Removing Stains and Cleaning Concrete Surfaces (IS214)

"Repair with Thin-Bonded Overlay," *Concrete Technology Today* (PL851)

Resurfacing Concrete Floors (IS144)

Slab Thickness Design for Industrial Concrete Floors on Grade (IS195)

Steel Fiber Reinforced Concrete Properties and Resurfacing Applications (RD049)

Subgrades and Subbases for Concrete Pavements (IS029)

Surface Treatments for Concrete Floors (IS147)

Transporting and Handling Concrete (IS178)

Understanding Concrete Floors and Moisture Issues (CD014)

White Cement Concrete (EB217)

To order, write, phone, or fax

Customer Service
Portland Cement Association
PO Box 726
Skokie, IL 60076-0726
847.966.6200, ext. 564
800.868.6733 toll free
847.966.9666 fax
Or visit our Web site at www.portcement.org.

Listings of PCA materials are available in the Association's various catalogs of publications, computer software, and audiovisual materials. Free.

PCA An organization of cement companies to improve and extend the uses of portland cement and concrete through market development, engineering, research, education, and public affairs work.

RELATED WEB SITES

American Concrete Institute
ACI www.aci-int.org

American Concrete Paving Association
ACPA www.pavement.com

American Society for Testing and Materials
ASTM www.astm.org

Canadian Standards Association
CSA www.csa.ca

Construction Specifications Institute
CSI www.csi.org

Decorative Concrete Council
DCC www.decorativeconcretecouncil.org/

Decorative Concrete Network
DCN www.decorative-concrete.net/

National Terrazzo and Mosaic Association
NTMA www.ntma.com

Portland Cement Association
PCA www.portcement.org

Post-Tensioning Institute
PTI www.post-tensioning.org